An Introduction to Source Coding

Raymond Veldhuis
Marcel Breeuwer

Prentice Hall
New York   London   Toronto   Sydney   Tokyo   Singapore

First published 1993 by
Prentice Hall International (UK) Ltd
Campus 400, Maylands Avenue
Hemel Hempstead
Hertfordshire, HP2 7EZ
A division of
Simon & Schuster International Group

© Prentice Hall International (UK) Ltd and Philips Laboratories, Eindhoven 1993

All rights reserved. No part of this publication may be reproduced, stored in a retrieval system, or transmitted, in any form, or by any means, electronic, mechanical, photocopying, recording or otherwise, without prior permission, in writing, from the publisher.
For permission within the United States of America contact Prentice Hall Inc., Englewood Cliffs, NJ 07632

Printed and bound in Great Britain by
Redwood Books Ltd, Trowbridge, Wiltshire

Library of Congress Cataloging-in-Publication Data

Veldhuis, Raymond.
    An introduction to source coding / Raymond Veldhuis, Marcel Breeuwer.
        p.   cm.
    Includes bibliographical references and index.
    ISBN 0-13-489089-2
    1. Signal processing—Digital techniques.   2. Coding theory.
I. Breeuwer, Marcel.   II. Title.
TK5102.5.V45   1993
621.382′2—dc20   93–25644
                                                                         CIP

British Library Cataloguing in Publication Data

A catalogue record for this book is available from the British Library

ISBN 0-13-489089-2

---

1 2 3 4 5   97 96 95 94 93

Contents

Figures ... ix

Tables ... xiii

Foreword ... xiv

Preface ... xv

Notations ... xviii
   List of symbols ... xxi
   List of abbreviations ... xxi

Part I Backgrounds ... 1

1 Source coding ... 3
  1.1 Introduction ... 3
  1.2 Source coding in a transmission system ... 3
  1.3 Source signals ... 5
  1.4 Bit rate and distortion ... 8
  1.5 Rate-distortion theory ... 12
  1.6 Source-coding aspects ... 14
  1.7 Blocks of a source-coding system ... 15
  1.8 Problems ... 17

2 The basic blocks of a coding system ... 19
  2.1 Introduction ... 19
  2.2 Coding, quantization and transformation ... 19
  2.3 Problems ... 30

3 Coding of discrete sources — 32
- 3.1 Introduction — 32
- 3.2 Entropy — 33
- 3.3 Codes — 34
- 3.4 Shannon's source-coding theorems — 36
- 3.5 Some coding methods — 39
- 3.6 Statistically dependent symbols — 45
- 3.7 Problems — 48

4 Quantization — 50
- 4.1 Introduction — 50
- 4.2 The quantizer as a discrete source — 51
- 4.3 The quantizer as an error source — 52
- 4.4 Types of quantizers — 56
- 4.5 Quantization and coding — 68
- 4.6 Relation with the rate-distortion theory — 70
- 4.7 Vector quantization — 72
- 4.8 Selecting a quantizer — 85
- 4.9 Problems — 86

5 Signals and transformations — 87
- 5.1 Introduction — 87
- 5.2 Quantizing and coding of dependent sources — 88
- 5.3 Relations to the rate-distortion theory — 93
- 5.4 Orthogonal transforms — 96
- 5.5 Subband filtering — 113
- 5.6 Predictive filtering — 119
- 5.7 Problems — 127

6 Adaptive source coding — 129
- 6.1 Introduction — 129
- 6.2 Forward and backward adaptivity — 130
- 6.3 Adaptation rate — 132
- 6.4 Parameter estimation — 133
- 6.5 Adaptation and coding delay — 137
- 6.6 Forward and backward adaptivity compared — 140
- 6.7 Adaptive transform coding — 140
- 6.8 Adaptive subband coding — 141
- 6.9 Adaptive linear predictive coding — 143
- 6.10 Bit-rate control — 144
- 6.11 Problems — 150

Part II	**Applications**	**153**

7	**Introduction to Part II**	**155**

8	**Subband coding of audio signals**	**156**
	8.1 Introduction	156
	8.2 Masking	157
	8.3 Subband coding	160
	8.4 A masking model for subband coding	162
	8.5 Adaptive bit allocation	165
	8.6 Results	168
	8.7 Other systems	168

9	**Picture source coding**	**169**
	9.1 Introduction	169
	9.2 Digital pictures	169
	9.3 The need for picture source coding	172
	9.4 Picture coding techniques	173

10	**Pulse-code modulation of pictures**	**174**
	10.1 Principle	174
	10.2 Example of adaptive PCM	175

11	**Differential PCM of pictures**	**181**
	11.1 Principle	181
	11.2 Prediction techniques	183
	11.3 Quantization of the prediction error	185
	11.4 Coding quantized prediction errors	187
	11.5 Examples of quantization and coding	187

12	**Transform coding of pictures**	**192**
	12.1 Introduction	192
	12.2 The transform	193
	12.3 Quantizing of transform coefficients	196
	12.4 Coding of quantized coefficients	205
	12.5 Specific applications	206

13	**Subband coding of pictures**	**211**
	13.1 Principle	211
	13.2 The filter bank	212
	13.3 Quantization and coding	217
	13.4 Subband and transform coding	222
	13.5 Specific applications	223

14 Pyramidal coding of pictures — 224
14.1 Principle — 224
14.2 Quantization — 227
14.3 Pyramidal and subband coding — 230
14.4 Specific applications — 231

15 Source coding of picture sequences — 232
15.1 Introduction — 232
15.2 Temporal DPCM — 233
15.3 Prediction techniques — 235
15.4 Quantization and coding — 237
15.5 Specific applications — 238

Part III Conclusions — 241

16 In conclusion to Part I — 243
16.1 Introduction — 243
16.2 Redundancy, irrelevancy and distortion — 244
16.3 Basic blocks of a source-coding system — 244

17 In conclusion to Part II — 247

18 Future developments — 249

References — 251

Index — 259

Figures

1.1	Transmission system	4
1.2	DPCM encoder and decoder	9
1.3	DPCM encoder with quantization	10
1.4	Example of the rate-distortion function for a real-value source	13
2.1	Example of a quantization characteristic	21
2.2	APCM source encoder	22
2.3	APCM source decoder	23
2.4	Simplified APCM encoder	23
2.5	Simplified APCM decoder	24
2.6	Blocks of a source-coding system	24
3.1	Code tree	35
3.2	Code tree	40
4.1	Additive quantization errors	52
4.2	Additive source-decoding model	53
4.3	Signal-to-noise ratios of the outputs of a 63-level and a 7-level quantizer as a function of the input variance	55
4.4	Example of a midtread quantization characteristic	56
4.5	Midrise quantization of small signals	59
4.6	Signal-to-noise ratios of a uniform quantizer and an A-law quantizer as a function of the input variance	63
4.7	Signal-to-noise ratios of a uniform quantizer and an A-law quantizer as a function of the input variance	64
4.8	Non-uniform quantizer and reconstructor, realized by means of expansion and compression functions	65
4.9	Signal-to-noise ratio of the output of an A-law quantizer as a function of the input variance	66
4.10	Construction of a non-uniform A-law companding quantization characteristic	67

Figures

4.11	Noise-shaping quantizer	68
4.12	Rate-distortion bound and Gish-Pierce asymptote for a Gaussian source	73
4.13	Encoder and decoder of full-search VQ	74
4.14	Pseudo-3D and top view of two-dimensional probability density function	76
4.15	One-dimensional probability density function	77
4.16	Scalar quantization levels and one-dimensional probability density function	77
4.17	Set of VQ codebook vectors and all combinations of scalar quantization levels	78
4.18	Encoder and decoder of a two-stage cascaded VQ	82
4.19	Gain-shape VQ encoder and decoder	84
4.20	Binary tree-structured VQ codebook	85
5.1	Source-coding system without transformation	87
5.2	Source-coding system with transformation	88
5.3	Power spectral density function of a second-order autoregressive process	95
5.4	Rate-distortion function and Gish-Pierce bound for an autoregressive process of order 2	96
5.5	Transform encoder with a block length of four	97
5.6	Transform decoder with a block length of four	98
5.7	Additive source-decoding model for transform coders	99
5.8	Contour plot of a joint Gaussian probability density function	101
5.9	Contour plot of a joint Gaussian probability density function after an orthogonal transform	102
5.10	Basic scheme of a subband source encoder and decoder	114
5.11	Blocks in a linear predictive source-coding system	120
5.12	Source encoder of a linear predictive coding system	122
5.13	Filters of a non-recursive linear predictive coding system	123
5.14	Example of a third-order filter	123
5.15	Source encoder of a non-recursive linear predictive coding system	124
5.16	DPCM encoder	124
5.17	Source encoder with feedback	124
5.18	Simplified CELP encoder without pitch prediction	127
6.1	Simplified diagram of a forward-adaptive source encoder	130
6.2	Simplified diagram of a forward-adaptive source decoder	131
6.3	Simplified diagram of a backward-adaptive source encoder	131
6.4	Simplified diagram of a backward-adaptive source decoder	131
6.5	Simplified diagram of a forward-adaptive source encoder	133
6.6	Simplified diagram of a backward-adaptive source encoder	133
6.7	Simplified diagram of a backward-adaptive source decoder	133
6.8	Non-recursive parameter estimation	136

Figures xi

6.9	Recursive parameter estimation	137
6.10	APCM encoder	138
6.11	Source encoder with buffer control	146
6.12	Writing to and reading from the buffer memory	146
6.13	Buffer control	147
6.14	Typical relation between step size and number of bits per block of samples	148
6.15	Simple subband source encoder and decoder with adaptive bit allocation	149
6.16	Source encoder with adaptive bit allocation	151
8.1	Threshold in quiet and masking thresholds of narrow-band noise maskers for tonal targets at various levels	158
8.2	Threshold in quiet and masking thresholds of narrow-band noise signals for tonal targets at various frequencies	159
8.3	Masking thresholds of critical-band noise targets with a 400 Hz masker at various sound-pressure levels	160
8.4	Subband-coding system with 32 subbands	162
8.5	Masking threshold as a function of frequency with subband boundaries	164
8.6	Computation of in-band masking level	164
8.7	Computation of masking level in subbands at higher frequencies	165
9.1	Rectangular sampling raster	170
9.2	Example of CCIR 601 TV frame	171
9.3	Basic block diagram of source encoder	173
10.1	Adaptive PCM encoder	175
10.2	APCM encoder	176
10.3	Division of a picture into blocks	176
10.4	Placement of quantization levels in APCM	177
10.5	APCM decoder	179
10.6	TV picture coded with APCM	179
10.7	APCM-coded picture before quantization, after subtraction minimum	180
10.8	Difference between original and APCM-coded picture	180
11.1	DPCM encoder	182
11.2	DPCM decoder	182
11.3	Order of treatment of samples in DPCM	184
11.4	DPCM prediction environment	184
11.5	Modelling of a non-uniform quantizer	186
11.6	Histogram of quantized DPCM prediction error (uniform quantization)	188
11.7	Histogram of quantized DPCM prediction error (non-uniform quantization)	190
11.8	Difference between original and DPCM-coded picture (non-uniform Q)	191

xii Figures

11.9 Difference between original and DPCM-coded picture (uniform Q)   191

12.1 Block diagram of transform coder   193
12.2 Basis matrices of DCT   195
12.3 Calculation DCT via DFT   196
12.4 Original block and its DCT coefficients   198
12.5 Illustration of the effect of quantization of a DCT coefficient   199
12.6 Block diagram of transform coding with weighting   200
12.7 Weighting function for DCT coding   200
12.8 Zonal coding with bit allocation   203
12.9 Adaptive zonal coding   203
12.10 Dead-zone quantizer   204
12.11 Zig-zag scanning of transform coefficients   205
12.12 Result of zig-zag scanning of transform coefficients   206
12.13 Mapping of quantized transform coefficients to codewords   208
12.14 Scanning of an interlaced TV picture   209
12.15 Motion-adaptive intraframe DCT coding   210
12.16 Division block of 8 by 8 samples into 2 of 4 by 8 samples   210

13.1 Basic scheme of a subband source encoder and decoder   212
13.2 Two-band subband filtering system   213
13.3 Quadrature-mirror filters   213
13.4 Impulse response and amplitude transfer function of subband filter   215
13.5 Overall subband filtering transfer function   216
13.6 Example three-band subband filtering system   216
13.7 Example of subband filtering of pictures   217
13.8 Division by subband filtering of spectrum of picture   218
13.9 Photograph of subband filtered picture   218
13.10 Subband filtering: splitting topologies   219
13.11 Quantization of lowpass subband signal   219

14.1 Basic pyramidal filtering scheme   225
14.2 Creation of a 'pyramid' of resolutions   226
14.3 Repeated pyramidal filtering   226
14.4 Pyramidal filtering: lowpass and highpass picture   228
14.5 Spectrum of quantization noise in pyramidal coding   229
14.6 Feedback of lowpass quantization errors in pyramidal coding   230

15.1 Differences between two successive TV frames   233
15.2 Histogram of differences between two successive TV frames   234
15.3 Temporal DPCM: encoder   234
15.4 Motion-compensated prediction in temporal DPCM   236
15.5 Displaced-frame differences   237
15.6 Temporal DPCM with motion-compensated prediction   238
15.7 Motion-compensated interframe DCT coding   238

Tables

3.1	Code for a source producing 2 symbols	41
3.2	Combined code for a source producing 2 symbols	41
4.1	Decision levels, probabilities and entropies of a midrise and a midtread quantizer	57
4.2	Bias of midrise and midtread quantizers with integer inputs	58
5.1	Bit rates and numbers of entries of Huffman tables for coders with and without an orthogonal transform	104
10.1	Example bit allocation table	177
11.1	Huffman code for DPCM	189
11.2	Non-uniform quantizer for DPCM	190
12.1	VLC and RLC tables for transform coefficient coding	207

Foreword

Over the last thirty or so years, digital signal processing hardware has become increasingly fast, small, efficient and economical, and last - but not least - inexpensive to buy. There seems to be no end to this trend in sight. Complex and sophisticated concepts, once considered to be only theoretical experiments, have been or are now being realized. One of these concepts, central to the efficient transmission and storage of information, is source coding.

In the first part of the book, the authors distinguish three consecutive operations in the general case of source coding. These are: transformation, quantization and lossless source coding. In this way, and by dealing with each of the three operations in depth, they succeeded in greatly clarifying the subject of source coding.

In the second part of the book, the theory outlined in the first part is applied to source coding of audio signals and pictures, both examples of great practical importance. What finally emerges is a coherent and lucidly written book, on which I would like to congratulate the authors. I trust that this book will strongly appeal to the scientific community for which it has been written.

Prof. dr. J.B.H. Peek,
Faculty of Mathematics and Computer Sciences,
University of Nijmegen, The Netherlands.

Preface

A growing proportion of transmission systems used in the world transmit digital signals. A digital telephony system is an example of such a digital transmission system. It can be expected that alongside the already existing analog transmission systems for radio and TV, digital ones will be introduced in the future. There are three reasons for the increasing interest in the transmission of digital signals. The first is that the quality that can be offered is better than in the analog cases. The second is that digital signals can be better protected against errors that corrupt them during transmission. The third reason is that additional signal processing before or after transmission is easier for digital than for analog signals.

A drawback of the use of digital transmission channels is their often limited capacity: the signal's source produces more bits per second than can be transmitted. This is why source-coding systems are often indispensable: they reduce the bit rate of the signal to one than can be transmitted. How that is done is the topic of this book. Its goal is to give a global understanding of source coding and to present an overview of practical coding schemes for speech, music and pictures.

Storage media, such as computer memories, digital tapes or optical disks, also exhibit the features of transmission channels. For instance, the frequency content of a signal must be adapted to a tape, errors are introduced by defects in computer memories, and memories may have a limited storage capacity. Storage media are therefore often treated as channels [1, 2, 3], and information and communication theory, dealing with transmission, are also applied to storage. It is therefore not surprising that source coding is also found in storage systems. The recently introduced digital compact cassette (DCC®) is an example of a digital consumer storage system in which source coding is crucial.

Part I of this book is mainly theoretical and provides background to source coding. It shows how source-coding systems can be divided into certain basic blocks, called transformation, quantization and coding blocks, each performing distinct functions. These blocks are discussed in separate chapters. In addition, the book contains a chapter on adaptivity in source-coding systems, which includes a discussion on bit-rate control. Part II discusses extensive examples of source-coding systems for sound signals and pictures. It relates the systems presented there to the theoretical concepts of Part I. Part III is a concluding part. It contains two chapters that summarize Parts I and II, and a chapter that discusses future developments. Part III is written in such a way that it can be read independently.

The approach whereby source-coding systems are divided into basic blocks which can be analysed separately distinguishes this book from others such as [4, 5]. This approach has the advantages that similarities between different coding systems such as transform, subband or predictive coders become evident and that these different systems can also be discussed within a unifying framework. A slight drawback of this approach is that the division into basic blocks is not always unique. Some of the operations performed in source-coding systems can be seen as belonging arbitrarily to one or other basic block. In those cases, although the division becomes somewhat arbitrary, it is no less useful.

The mathematical theory of source coding is rate-distortion theory [6, 7]. Rate-distortion theory gives formal bounds to the bit rates needed to transmit signals at a given distortion. However, it rarely gives practical methods of achieving these bounds. An exception is the vector quantizer discussed in Chapter 4. In contrast to rate-distortion theory, which is the mathematical theory of source coding, this book tries to present a theoretical background to source-coding systems. Practical source-coding systems are almost never optimal in terms of rate-distortion theory. Nevertheless, rate-distortion theory is used to explain why the basic blocks in a source-coding system are present and how they help to reduce the bit rate.

Source coding as discussed here is a broad discipline: it involves statistics, digital signal processing, signal theory, information theory and the theory of human perception. This makes it an interesting but at the same time a difficult discipline. Since the book aims to be an introduction rather than a reference work, it would be too much to expect an understanding of all these topics. However, to be able to appreciate the contents of Part I in particular, the reader should be familiar with the basics of digital signal processing, as presented in, for example, [8, 9], those of the theory of stochastic signals as in, say, [10], and the basics of linear algebra as in, say, [11]. For reference purposes [4] and [5] are recommended. Some of the derivations of important results are tedious. It is beyond the scope of this book to present them fully. Other derivations are necessary or illustrative and are not omitted. In many cases the results of Part I are introduced by means of examples which, therefore, cannot be ignored. The examples are typeset in a smaller character font and they end with ◊. A few exercises have been added at the end of the chapters.

The book aims to be a textbook that can be used in a senior-level course in advanced digital signal processing or in communications. It is also suitable for advanced junior-level students and first-year graduate students. In addition it contains useful information for communication engineers and those starting as research scientists.

We would like to thank our colleagues from the digital signal processing group at Philips Research Laboratories in Eindhoven for the stimulating discussions about and their continual interest in the contents of this book while it was being written.

Special thanks go to Rob Beuker, Richard Heusdens, Werner Oomen and Robbert van der Waal for reviewing the manuscript. The management of Philips Research Laboratories in Eindhoven is gratefully acknowledged for giving us the opportunity to write this book.

Notations

The following paragraphs present the notations, symbols and abbreviations that are used throughout the book. In general, notations and symbols are explained when they are used for the first time, nevertheless the following pages may serve as a reference. First general notations for scalars, vectors, etc. are given, then specific symbols that are always used to denote the same entity are listed. Finally, a list of abbreviations is given.

The symbols $\mathbb{N}$, $\mathbb{Z}$, $\mathbb{R}$ and $\mathbb{C}$ denote, respectively, the sets of the natural numbers, integers, real numbers and complex numbers. The lower and upper case italic roman and Greek letters are reserved for scalars. The lower case bold face roman letters are reserved for vectors and the upper case bold face roman letters are reserved for matrices. For example, a, B, γ and Δ are scalars, $\mathbf{e}$ is a vector and $\mathbf{D}$ is a matrix. The vector $\mathbf{e}$ of length n has elements

$$e_i, \ i = 1, \ldots, n,$$

or

$$e_i, \ i = 0, \ldots, n-1,$$

whichever notation is convenient. The $m \times n$ matrix $\mathbf{D}$ has elements

$$d_{i,j}, \ i = 1, \ldots, m, \ j = 1, \ldots, n,$$

or

$$d_{i,j}, \ i = 0, \ldots, m-1, \ j = 0, \ldots, n-1,$$

whichever notation is convenient.

Vector and matrix transposition is denoted by the superscript T, as in $\mathbf{e}^\mathrm{T}$ and $\mathbf{D}^\mathrm{T}$. The 2-norm [12] of a vector or matrix is denoted by $\|\cdot\|$. The inverse of the matrix $\mathbf{M}$ is $\mathbf{M}^{-1}$. The determinant of $\mathbf{M}$ is $|\mathbf{M}|$. The inverse of the matrix $\mathbf{M}^\mathrm{T}$ is $\mathbf{M}^{-\mathrm{T}}$. The symbol $\mathbf{I}$ is used for the identity matrix, $\mathbf{0}$ is an all-zero vector or matrix of appropriate sizes.

The order of magnitude of a number q is denoted by $O(q)$. The function $\delta(x)$, $x \in \mathbb{R}$, denotes the Dirac delta function; δ_k, $k \in \mathbb{Z}$, is the Kronecker delta function.

A sampled data sequence is denoted by a lower case roman letter followed by a time index between square brackets. For instance $s[i]$ is the ith sample of the sampled data sequence

$$s[j],\ j = -\infty, \ldots, +\infty,$$

and

$$s[k],\ k = 0, \ldots, N-1,$$

is a segment, or sub-sequence, taken from this sequence. If the sequence character is emphasized, the notations

$$\{s[k]\}$$

and

$$\{s[k]\}_{k=0,\ldots,N-1}$$

are used for the entire sequence or the sub-sequence respectively.

A picture is a two-dimensional sampled data sequence denoted by

$$p[m,n],\ m = 0, \ldots, M-1, n = 0, \ldots, N-1.$$

In this notation m is the vertical and n the horizontal coordinate. This means that M is the number of lines in the picture and N the number of samples on a line. The samples in a picture are usually referred to as pixels or pels. A picture sequence is denoted as

$$p_i[m,n],\ m = 0, \ldots, M-1, n = 0, \ldots, N-1,$$

where i is the time index.

The impulse response of a filter is either denoted as a segment of data, for instance,

$$h[l],\ l = 0, \ldots, q,$$

is possibly the impulse response of a finite impulse response filter of order q and length $q+1$, or as lower case roman letters with a subscript. The above impulse response is then denoted as

$$h_l,\ l = 0, \ldots, q.$$

The z-transform [8] of sequences and impulse responses is defined by

$$S(z) = \sum_n s[n] z^{-n},$$

and the Fourier transform by

$$S(\exp(\mathrm{j}\theta)) = \sum_n s[n] \exp(-\mathrm{j}\theta n).$$

For cosmetic reasons, $\exp(\mathrm{j}\theta)$ is sometimes replaced by $\mathrm{e}^{\mathrm{j}\theta}$.

In the figures a box:

xx Notations

$$s[n] \rightarrow \boxed{z^{-1}} \rightarrow s[n-1]$$

represents a unit-delay element. A box:

$$s[n] \rightarrow \boxed{\downarrow L} \rightarrow s[nL]$$

represents a decimator. Finally, a box

$$s[n] \rightarrow \boxed{\uparrow L} \rightarrow y[n] = \begin{cases} s[k], & n = kL, \\ 0, & \text{otherwise,} \end{cases}$$

represents an interpolator.

Especially in Part I scalars, vectors, matrices and signal values can be stochastic variables. Usually this is clear from the context and no special notation is used to distinguish stochastic variables from realizations.

The probability density functions of the stochastic variable y and of the stochastic vector $\mathbf{u}$ are $p_y(\cdot)$ and $p_\mathbf{u}(\cdot)$. The statistical expectation operator is denoted by $\mathcal{E}\{\cdot\}$. The expected value, also called the mean, of stochastic variable y is denoted by $\bar{y}$, the variance of y is denoted by σ^2, or σ_y^2. The subscripts are used only if confusion would otherwise occur. In the literature different definitions exist for autocorrelation and autocovariance functions [10, 13]. Here the term autocovariance function is not used. The autocorrelation function of the stationary stochastic signal s_i, $i = -\infty, \ldots, +\infty$, is denoted by $R(k)$, or $R_{ss}(k)$ and defined by

$$R(k) = \mathcal{E}\{(s_i - \mu)(s_{i+k} - \mu)\}, \quad k = -\infty, \ldots, +\infty.$$

In most of the cases discussed in this book, μ is assumed to be equal to zero and

$$R(k) = \mathcal{E}\{s_i s_{i+k}\}.$$

The power spectral density functions of the signal s_i, $i = -\infty, \ldots, +\infty$, is denoted by $S(\theta)$, or $S_{ss}(\theta)$ and defined by [13]

$$S(\theta) = \sum_{k=-\infty}^{\infty} R(k) \exp(-j\theta k), \quad -\pi \leq \theta \leq \pi.$$

Estimates or approximations for samples or parameters are denoted with a circumflex. For instance, $\hat{\mathbf{a}}$ is the estimate for the vector $\mathbf{a}$.

List of symbols

The following list summarizes the specific symbols used throughout this book. If a symbol is not found in this list, it is probably described in the first part of this section explaining the more general notations. It is sometimes necessary to introduce variation by adding subscripts or superscripts to these symbols, especially in the case of pictures where sequences are two or three dimensional.

$x[i]$	input signal to a source encoder
$\hat{x}[i]$	output signal from a source decoder
$y[i], \mathbf{y}, y_i$	input to a quantizer
$\hat{y}[i], \hat{\mathbf{y}}, \hat{y}_i$	output from a reconstructor
$q[i]$	output sequence from a quantizer
$e[i]$	source-coding error
N	length of a segment of samples, coded or quantized together
M	number of distinct quantizer-output symbols
$\{a_1, \ldots, a_M\}$	set of quantizer-output symbols
$\{l_1, \ldots, l_M\}$	set of quantization levels
$\{t_0, \ldots, t_M\}$	set of decision thresholds
$H(\cdot)$	entropy
$H(\cdot|\cdot)$	conditional entropy
$h(\cdot)$	differential entropy
$\overline{s}$	average codeword length
$\overline{s}^{(n)}$	average codeword length for a group of N symbols
D	distortion
R	bit rate
$R_\mathrm{B}(D)$	rate-distortion bound
Δ	step size of a uniform quantizer
$R(k),\ k = -\infty, \ldots, +\infty$	signal's autocorrelation function
$S(\theta),\ |\theta| \leq \pi$	signal's spectrum
$\mathbf{R}$	signal's $N \times N$ autocorrelation matrix
$\mathbf{C}$	$N \times N$ correlation matrix of two vectors

List of abbreviations

ADPCM	adaptive differential pulse-code modulation
ADRC	adaptive dynamic range coding
APCM	adaptive pulse-code modulation
CCIR	Comité Consultatif International de Radio
CELP	codebook excited linear prediction

CQF	conjugate quadrature filters
DCC	digital compact cassette
DCT	discrete cosine transform
DFD	displaced-frame difference
DFT	discrete Fourier transform
DPCM	differential pulse-code modulation
EOB	end-of-block
ESP	electronic still picture
FLC	fixed-length coding
HDTV	high-definition television
ISO	International Standardization Organization
JPEG	Joint Picture Experts Group
KLT	Karhunen-Loève transform
LBG algorithm	Linde-Buzo-Gray algorithm
LPC	linear predictive coding
MPEG	Motion Picture Experts Group
MSE	mean-square error
PASC	precision adaptive subband coding
PC	pyramidal coding
PCM	pulse-code modulation
QMF	quadrature-mirror filters
SBC	subband coding
SDTV	standard-definition television
SNR	signal-to-noise ratio
SPL	sound-pressure level
TC	transform coding
TV	television
VAPCM	vector-adaptive pulse-code modulation
VLC	variable-length coding
VQ	vector quantization
WHT	Walsh-Hadamard transform

Part I

Backgrounds

1
Source coding

1.1 Introduction

Part I of this book gives theoretical backgrounds to source coding. It shows how any source-coding algorithm can be divided into a number of basic functional blocks. Before this is done in Chapters 2 to 6, Sections 1.2 to 1.6 of this chapter briefly introduce the subject of source coding and explain why and how source-coding systems are used. Section 1.7 summarizes Part I of this book.

1.2 Source coding in a transmission system

Figure 1.1 shows a schematic diagram of a transmission system such as can be found as the first figure in almost any textbook on information or communication theory. It contains most of the components of a transmission system.

A transmission system transmits signals from a source to a destination, which is sometimes called a sink [4]. The source is the device by which the signals are (thought to be) generated. The destination is the place where they are received.

The central part of the transmission system shown in Figure 1.1 is the channel. This is the medium which physically transports the signal, for example as an electrical signal, as an electromagnetic wave or as a sequence of light pulses. Practical examples of channels are a coaxial cable, an optical fiber and the atmosphere. Apart from the channel, the signal passes through three pairs of blocks: the modulator and demodulator blocks, the error protection and correction blocks, and the source encoder and decoder blocks. These pairs of blocks are present because of three channel properties discussed below.

The first channel property is that channels can only transport physical (e.g. electrical) signals. For successful transmission a digital signal must therefore be converted to a form appropriate for the channel. For radio transmission, for instance, the signal must be modulated on a carrier and then converted into an electromagnetic wave. The modulator converts the signal to a representation that

4 Source coding

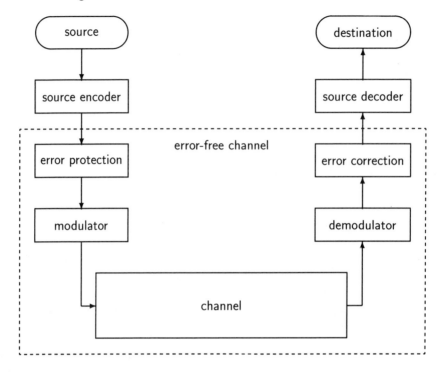

Figure 1.1 *Transmission system.*

is appropriate for the channel. The demodulator converts the received signal back to its original form. For basic literature on modulation principles the reader is referred to [14, 15].

The second channel property is that, even if a signal is adapted to the channel, it does not always pass it undisturbed: the channel may introduce errors. If digital information is transmitted, the result can be that after demodulation some bits are inverted. The block before the modulator is the error-protection unit. It adds bits to the signal that make it possible for the error-correction unit to detect and even correct errors introduced by the channel. For basic literature on error correction the reader is referred to [16, 17].

The third channel property is that there is an upper bound to the number of bits per second that can be correctly transmitted. This bound is called the channel capacity. The source-encoding block reduces the number of bits per second with which the input signal is represented, to a number that is low enough for transmission. The signal with the reduced bit rate is the source-encoded signal. The source decoder converts this to a reconstruction of the input signal. Unfortunately, source encoding and decoding may change the signal. This results in the reception of a distorted signal. In a good source-coding system the distortion is kept below a certain level.

It is mainly source coding for speech, music and pictures that is considered here. This implies that one finds a human observer at the destination. This has its impact on the notion of distortion and on the design of source-coding systems. As has already been remarked, a good source-coding system keeps distortion below a certain level. If signals such as speech, music and pictures are received by a human observer, it means that after reception these signals must have a desired subjective quality rather than a desired objective quality.

In most of what follows it is assumed that the concatenation of the error-protection block, the modulator, the channel, the demodulator and the error-correction block behaves as a digital, error-free channel, which implies that the source decoder receives the undistorted output of the source encoder.

Source coding is not only the name for the discipline involved with the design of source-coding algorithms and systems but also for the action of the source encoder and decoder. Other names that are sometimes used for source coding are bit-rate reduction, data reduction and data compression. The combination of a source encoder and decoder is often called a codec.

Examples of source coding applied in transmission and storage systems are: source coding of speech signals in mobile automatic telephony, source coding of x-ray and nuclear magnetic resonance images for storage in medical databases, source coding of sound signals for storage on compact disc interactive (CD-I®) disks and on digital compact cassette (DCC®) tapes and for digital audio broadcasting, source coding of images of documents for storage in the Megadoc system, and source coding of digital TV pictures for storage on a digital video tape.

1.3 Source signals

Before the aspects of source coding are discussed further, some remarks must be made on possible signal sources. Source signals are often classified into the five categories mentioned below.

Real-value continuous-time signals

From a mathematical point of view, these signals are functions of time. They are, for instance, denoted as

$$x(t),$$

with $t \in \mathbb{R}$ and $x(t) \in \mathbb{R}$. The parameter t denotes time. These signals are usually called analog signals.

6 Source coding

Real-value discrete-time signals

From a mathematical point of view, real-value discrete-time signals are sequences of real numbers. A number is called a sample. The nth sample, for instance, is denoted as

$$x[n],$$

with $n \in \mathbb{Z}$ and $x[n] \in \mathbb{R}$. The parameter n represents time; it corresponds to an instant nT_s, where T_s is the sampling period. A sequence of numbers is usually also denoted as $x[n]$, and a sub-sequence, for instance, as $x[n]$, $n = 0, \ldots, N-1$. If the sequence character has to be emphasized, the notations

$$\{x[n]\}$$

and

$$\{x[n]\}_{n=0,\ldots,N-1}$$

are used for the entire sequence and a sub-sequence, respectively.

A real-value discrete-time signal can be derived from a real-value continuous-time signal by a process called sampling. In that case

$$x[n] = x(nT_s).$$

If the highest frequency that occurs in $x(t)$ is less than $1/(2T_s)$, then the real-value continuous-time signal can be perfectly reconstructed from the sequence $\{x[n]\}$ [8]. The sampling frequency or sampling rate f_s is defined by

$$f_s = \frac{1}{T_s}.$$

Signals occurring in charge-coupled devices (CCDs) are examples of real-value discrete-time signals.

Integer-value discrete-time signals

From a mathematical point of view, integer-value discrete-time signals are sequences of integer numbers. This is the class of signals that are generally referred to as digital signals. The same notations as for real-value discrete-time signals are used, but now $x[n] \in \mathbb{Z}$. If they have a finite range, they are suited for digital signal processing, since their values can be represented digitally with a finite number of bits. For instance, if

$$-M \leq x[n] < M - 1$$

then the samples can be uniquely represented digitally by $\lceil \log_2(2M) \rceil$ bits. The notation $\lceil a \rceil$ means the first integer that is greater than or equal to a.

Discrete-value discrete-time signals

From a mathematical point of view, discrete-value discrete-time signals are sequences of abstract symbols $q[i]$. The symbols are elements from a set, for example

$$q[i] \in \{a_1, \ldots, a_m\}.$$

The elements of a set containing m symbols can be uniquely represented digitally by $\lceil \log_2(m) \rceil$ bits. The class of integer-value discrete-time signals is a subclass of the discrete-value discrete-time signals. Sequences of ASCII characters also belong to this class.

Discrete-value continuous-time signals

An example of a discrete-value continuous-time signal is the random asynchronous telegraph signal [10], which is a continuous-time signal that can only attain two values, for example

$$x(t) \in \{0 \text{ V}, 5 \text{ V}\}.$$

Discrete-value, continuous-time signals are often used to model signals in a physical transmission channel.

Only one-dimensional signals have been mentioned. Multi-dimensional signals are also of interest, as this book also deals with source coding of pictures. Extension of the five categories above to more dimensions is easily done. A picture, for instance, is a two-dimensional function of space. If the space coordinates are discrete, then a picture sample may be denoted as

$$p[m, n].$$

Chapter 9 in Part II gives a more extensive introduction to multi-dimensional sequences.

This book mainly considers source coding of physical signals, such as speech, music and pictures. The signal processing in source coding is digital. This means that actually only the class of integer-value discrete-time signals is acceptable as input to a source encoder. Physical signals often have another representation. Music or speech signals at the output of a microphone, for instance, are continuous-time electrical signals. These signals have to be converted by a so-called analog-to-digital converter [8]. If a real-value continuous-time signal is desired at the destination, then a digital-to-analog converter has to be used to convert the source decoder output to an analog signal. Analog-to-digital conversion can be seen as a two-step operation. The real-value continuous-time signal is converted by sampling to a real-value discrete-time signal, which is subsequently rounded to a integer-value continuous-time signal.

Real-value discrete-time signals are sometimes mistakenly called digital signals. Even though practical source coding is always done on integer-value discrete-time signals it will be often assumed in this book that the signals are real-value and continuous-time. The reason is that it is sometimes easier to analyse this type of signal, and to correct the results afterwards, than to analyse digital signals directly.

Discrete-value discrete-time signals also play an important role in this book. This is because some of the most important theoretical results in source coding that are also of value for physical signals are derived for sources producing abstract symbols.

1.4 Bit rate and distortion

Let us assume that the source of Figure 1.1 produces digital signal samples at a rate of f_s samples per second, and that each sample is represented with R_s bits. The number R_s is the bit rate of the source signal, expressed in bits per sample. The product

$$I_s = f_s R_s \tag{1.1}$$

gives the bit rate of the source signal in bits per second. Both R_s and I_s are used to express bit rates. As has already been mentioned in Section 1.2, the number of bits per second that can be transmitted through a channel is limited. We also assume that the digital error-free channel of Figure 1.1 can transmit a number of I_c bits per second, which is less than I_s. This number I_c is called the transmission rate. A source coder reduces the bit rate of a source signal to the transmission rate.

In fact, there are two possibilities for the transmission rate. One is that it is fixed, which is the case in many broadcasting, transmission and storage applications. The other is that it is variable, which is, for example, the case for packet-switched networks or for storage on computer disks. In the case of a variable bit rate there may be restrictions on the average and the peak bit rates.

It is not obvious that bit-rate reduction of a signal leads to signals of acceptable quality after source decoding. There are two reasons why source coding can be feasible at all. They are discussed below.

The first reason is that many signals have a higher bit rate than is necessary for an error-free representation. In terms of the information theory: such a signal's representation is redundant. A reduction of the bit rate is then obtained by using another representation with fewer bits per second. It is said that in this case source coding removes the redundancy. Redundancy will be defined more formally later on. Removing the redundancy is a reversible process: the source decoder can always recover the original input signal. This type of source coding is sometimes called lossless or errorless or error-free source coding.

The second reason is that a perfect reconstruction of the input signal is not always required after source decoding, but that differences up to a certain degree between the input and the decoded signal are either acceptable or imperceptible. If

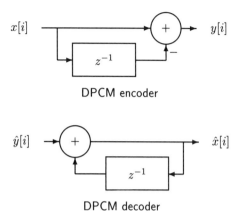

Figure 1.2 *DPCM encoder and decoder.*

errors after source decoding are allowed to some extent, the bit rate can be reduced by finding a less accurate representation with fewer bits per second, which no longer allows a perfect reconstruction of the input signal by the source decoder. The fact that source encoder input and source decoder output are different is irrelevant to the observer. An encoder working in this way is said to remove irrelevancy from the signal's representation. This is irreversible. The amount of deviation between the input signal and the decoded signal is called the distortion.

Example 1 Consider as an example of a redundant signal a monochrome digital TV picture. This is a matrix of samples, called picture elements or pixels, each having a value in the range $[0, 255]$. The pixels contain the luminance information of the picture. Pixels equal to zero represent black spots, pixels equal to 255 represent white spots. A pixel having a value between zero and 255 represents a grey spot. Each pixel is represented with eight bits. Assume, in this example, that the pixel values on a horizontal line do not change much over parts of the line, for example the absolute difference is always less than 7. This implies that the differences can be represented in four bits. A reduction of the bit rate is possible by first transmitting the first pixel of a line and subsequently all the differences between adjacent pixels on that line. The bit rate is thus reduced to about four bits per pixel. The source decoder can reconstruct the picture perfectly in the following way. The first pixel of a line is received uncoded. The second pixel can be obtained by adding to the first pixel the transmitted difference between the second and the first pixel. The third pixel can be obtained by adding the transmitted difference between the third and second pixel to the decoded second pixel, and so on. This technique is a simple version of what is generally known as differential pulse-code modulation or *DPCM* [4]. A block diagram is shown in Figure 1.2. The z^{-1} in this figure is the z-transform of a unit sample delay. The box labelled z^{-1} performs such a delay.

◇

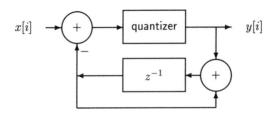

Figure 1.3 *DPCM encoder with quantization.*

Example 2 Example 1 can be extended slightly to illustrate that a further reduction of the bit rate is possible if some distortion in the received signal can be allowed. Suppose that in the decoded picture a lower resolution is acceptable, such that only the even values in the range $[0, 255]$ have to be represented. In that case, a further reduction of 1 bit per pixel can be achieved by a source encoder as depicted in Figure 1.3.

Careful inspection of Figure 1.3 shows that the pixel differences in the source encoder are taken between the input pixel and a delayed decoded output pixel. The quantizer maps the pixel differences on to the even values in the range $[-7, 7]$. The number of even values in this range is 7, which means that each value can be coded with 3 bits. This coding part is not shown in the diagram.

The source decoder can reconstruct the picture with an acceptable distortion using the same procedure as in Example 1.

◇

Example 2 shows that source coding may introduce distortion. In fact, that will be the case for most of the source-coding methods discussed here. The amount of distortion depends on the signal, bit rate, and source-coding algorithm. Distortion and bit rate play a role in the selection of a source-coding method. A problem is that a good mathematical measure for distortion that reflects perceptual quality is often hard to find. For audio, video and speech signals, for instance, the distortion measure should express the perceived loss in quality, but often it does not. The most general expression for a distortion measure is

$$D = f(\{\hat{x}[i]\}, \{x[i]\}), \tag{1.2}$$

where $\{x[i]\}$ and $\{\hat{x}[i]\}$ denote the entire input and output sequences of the source encoder and decoder. If, for two source-coding schemes 1 and 2, we have $D_1 > D_2$, this should mean that the output of codec 2 should be of higher perceptual quality than the output of codec 1.

The input signal for a source-coding system is not known in advance but knowledge of the signal is often available in statistical terms, such as probability density functions and power spectral densities. In the design and analysis of source-coding systems therefore, a statistical approach is usually adopted. This will usually also be the case in this book. Knowledge of the signal's statistical properties is often referred to as a statistical signal model. Knowledge of the observer is sometimes called an observer model. A statistical distortion measure has the form

$$D = \mathcal{E}\{f(\{\hat{x}[i]\}, \{x[i]\})\}, \tag{1.3}$$

where $\mathcal{E}\{\}$ denotes statistical expectation.

Practical distortion measures that are often used, but mostly fail to express perceived loss of quality, are the mean-square error (MSE)

$$D_{\text{MSE}} = \mathcal{E}\{(\hat{x}[i] - x[i])^2\} \tag{1.4}$$

and the signal-to-noise ratio (SNR)

$$D_{\text{SNR}} = \frac{\mathcal{E}\{(x[i])^2\}}{\mathcal{E}\{(\hat{x}[i] - x[i])^2\}}. \tag{1.5}$$

The first is an estimate of the power of the source-coding error. The second is the signal power divided by the mean-square error. Note that these distortion measures are sample-based, which is less general than (1.3). In both (1.4) and (1.5) it is assumed that $x[i]$ and $\hat{x}[i]$ are wide-sense stationary [10], which implies that D_{MSE} and D_{SNR} are independent of time index i.

That the practical measures (1.4) and (1.5) only have a limited usefulness is easily seen if one considers a speech-coding scheme which produces an almost perfect reconstruction of the input that is delayed for a certain number of samples. Both (1.4) and (1.5) will yield large distortions, but the error will be inaudible.

A distinction can be made between two source-coding principles. The first is waveform coding, which is the subject of this book; the second is parameter coding of signals.

In waveform coding the objective is to find a representation of the signal such that after source decoding a good approximation of the original waveform remains. What is good in this context depends on the distortion measure. Since the output sequence $\{\hat{x}[i]\}$ tries to approximate the input sequence $\{x[i]\}$, a distortion measure based on the difference between output and input sequences

$$D = \mathcal{E}\{f(\{\hat{x}[i]\} - \{x[i]\})\}, \tag{1.6}$$

seems appropriate for waveform coders. The mean-square error (1.4) is an example of such a distortion measure. For analysing waveform coders it is useful to define an additive source-coding error

$$e[i] = \hat{x}[i] - x[i]. \tag{1.7}$$

If a source-coding system introduces a delay, the distortion measure (1.7) will be large just because of the delay. In such cases it is better to define the source-coding error by

$$e[i] = \hat{x}[i] - x[i - L], \tag{1.8}$$

where L is the delay expressed in sample periods. The distortion measure

$$D = \mathcal{E}\{f(e[i])\} \tag{1.9}$$

can be used with either (1.7) or (1.8), whichever is more appropriate.

In parameter coding it is assumed that a generation model of the signal is known. Roughly speaking, a generation model is a mathematical procedure that generates the entire signal from a limited set of model parameters. The objective is to extract these model parameters from the signal, code them at a low bit rate and transmit them. The decoder tries to regenerate the signal from the received parameters. A good approximation of the waveform may not be obtained, but the result could be a signal sounding or looking similar to the original signal. Examples of parameter coders are vocoders for speech signals [18] and fractal coders for pictures [19]. The bit rates for parameter coding can be far lower than for waveform coding. However the quality after decoding is usually also lower. Pure waveform coding and pure parameter coding are two extremes. In many waveform coders signal models are also used and parameters are coded.

Some confusion may arise concerning signal models. So far, both statistical signal models and generation models have been mentioned as signal models. The distinction between them is that a statistical model is a collection of statistical properties of the signal whereas a generation model is a procedure to generate a signal starting from a set of parameters. In this book, the term signal model always refers to a statistical signal model.

1.5 Rate-distortion theory

Once a distortion measure has been chosen, it is permissible to ask whether for a given source an optimal source-coding method can be found that achieves a minimal bit rate at a given distortion. The rate-distortion theory [6, 7] states that under certain assumptions there exists a bound giving the minimum rate required for a signal at a given distortion. It also states that by coding long sequences of samples this bound can be approximated arbitrarily closely. This bound is called the rate-distortion function or rate-distortion bound $R_B(D)$, where D is the distortion according to the chosen distortion measure. The derivation of this bound is, unfortunately, in many practical situations an unsolvable mathematical problem. Moreover, a source-coding method trying to closely approach this bound may be of an extremely high complexity. For these reasons and for the lack of a good distortion measure, the rate-distortion theory is not directly of practical value for designing source-coding methods. It may, however, give some theoretical insights. In this book it will be used to explain certain concepts and principals.

Rate-distortion functions can be derived for both real-value and discrete-value sources. Figure 1.4 shows a rate-distortion function for a source producing real-value

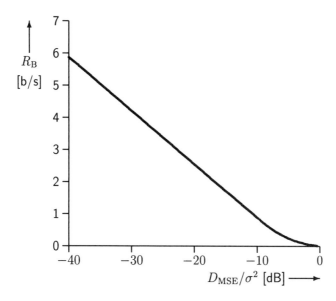

Figure 1.4 *Example of the rate-distortion function for a real-value source.*

samples with variance σ^2. The distortion measure is D_{MSE}/σ^2. The distortion axis ranges up to 0 dB, since mean-square errors greater than the signal variance can be avoided by simply transmitting zeros.

The rate-distortion function for a discrete-value source differs from that of a real-value source. Since the samples only have finite precision, errorless coding, as discussed in Chapter 3, can be used to reduce the bit rate by removing redundancy. The bit rate that can be achieved in this way determines the intersection of the rate-distortion bound with the rate axis. The rate-distortion bound for real-value sources always intersects the rate axis at $R = \infty$. Figure 1.4 shows a rate-distortion bound for a real-value source and the rate-distortion bound seems to intersect the rate axis. The diagram is, however, misleading, because the distortion axis is logarithmic.

Rate-distortion functions have a number of properties that are easy to express but sometimes hard to prove [6, 7]. For instance, they are convex, which means that cords are always above the rate-distortion function, and they are continuous for $D > 0$. Theoretically, any point (D, R) above the rate-distortion function can be achieved with an appropriate source-coding algorithm.

Practical source-coding algorithms are often suboptimal: at a given distortion their rate is greater than $R(D)$. In Chapters 4 and 5 the performances of some source-coding methods are compared with the rate-distortion function.

It can be concluded from the preceding paragraphs that a good source-coding system is based on knowledge of the signal as well as on knowledge of the observer. Knowledge of the observer is needed to select the appropriate distortion measure

14   Source coding

and to determine to what extent source-coding errors can be introduced. This specifies a certain point on the distortion axis. Knowledge of the signal is required to achieve the desired bit rate at this distortion. It seems desirable to choose a rate that is close to $R_B(D)$ for a given distortion D. However, $R_B(D)$ is often unknown and, even if that is not the case, this approach can result in a complex source coder with a long delay. Complexity, delay and other aspects that play a role when designing a source-coding system are briefly summarized in Section 1.6.

1.6 Source-coding aspects

In this section aspects which play a role when a source-coding method is selected or designed are summarized. Bit rate and distortion have already been mentioned and will only be discussed in relation to the other aspects.

Delay

Source coders often introduce a coding delay. For example, the performance of the simple source coders for pictures of Examples 1 and 2 can be improved as follows at the expense of introducing a coding delay.

Example 3 The distortion introduced by the source encoder of Example 2 can be reduced at the expense of coding delay and a minor increase in bit rate by storing a picture line first and calculating the pixel differences. If these are in the range $[-3, +3]$, the source encoder of Example 1 is used, otherwise the source encoder of Example 2 is used. One extra bit per line is transmitted to tell the source decoder how to decode the received data.
◇

It depends on the application to what extent coding delay is acceptable. In a telephony system, for instance, less delay can be accepted than in a recording or broadcasting system. As is the case in Example 3, delay is often introduced because the signal is stored and analysed before source encoding.

Complexity

Another important aspect of a source-coding scheme is its complexity. Complexity is often expressed in the number of operations per sample period or, if the source-coding system is to be realized as an integrated circuit, in chip area. Source-coding systems that achieve a low bit rate at a high quality can be very complex. The complexity can be reduced if a lower quality or a higher bit rate is acceptable. Again, it depends on the application as to whether a high complexity is problematic. For instance, a higher complexity can often be more acceptable in professional systems than in consumer products. In a broadcasting system there are usually just a few source encoders, which are pieces of professional equipment. They are therefore allowed to be far more complex and expensive than the source decoders, which are parts of consumer products.

Sensitivity to model deviations

A source-coding algorithm is usually based on, often statistical, assumptions about the signal. For instance, it is assumed in Examples 1 and 2 that the pixel differences are in the range $[-7, +7]$. Problems may occur if the signal does not satisfy these assumptions. A source-coding algorithm is insensitive to model deviations if it still works reasonably well when the signal deviates from the model. Such a source-coding method is often preferred if the nature of the signals that will be transmitted in the future is not known.

Example 4 The source-coding system of Example 3 can be rendered more robust by making the quantizer dependent on the maximum pixel difference in a line. If the pixel differences are in the range $[-3, +3]$, the encoder of Example 1 is used; if they are in the range $[-7, +7]$, the quantizer maps them on to the multiples of 2 in that range; if they are in the range $[-15, +15]$, they are mapped on to the multiples of 4 in the same range, and so on. Some bits must be used to tell the receiver which quantizer has been used, and this slightly increases the bit rate. The advantage is that a larger class of pictures can be coded with this system.
◇

Sensitivity to channel errors

Sometimes channel errors are neither corrected nor detected. In that case the source decoder receives erroneous information and additional distortion is introduced. Some source-coding methods are more sensitive to channel errors than others. Usually, the sensitivity to undetected errors increases with decreasing bit rate. If channel errors are detected but cannot be corrected then it is sometimes possible to conceal the errors after source decoding [20].

1.7 Blocks of a source-coding system

Source-coding systems can be divided into a number of basic blocks. Two of these blocks, quantization and transformation blocks, can already be identified in the simple system of Example 2. The other basic blocks are coding and control blocks. It will become clear that the basic blocks perform distinct functions that are elementary to source coding. All the basic blocks will be treated thoroughly in Chapters 3–6, but are briefly introduced below and in somewhat greater detail in Chapter 2.

Transformation blocks

Transformation blocks are used to obtain a signal representation that is suited to source coding. In the source encoder of Example 2 the transformation consists of subtracting the delayed encoded output pixel from each input pixel. An 'inverse' transformation is present in the source decoder[1]. Transformation blocks are discussed in Chapter 5.

Quantization blocks

Quantization blocks reduce the accuracy of the representation of the output of the transformation block. In this way they remove irrelevancy and introduce source-coding errors. The source encoder of Example 2 contains a quantizer. Quantization blocks and their counterparts in the source decoder, which are called reconstruction blocks, are discussed in Chapter 4

Coding blocks

Coding blocks have not occurred in the examples so far. They have two tasks: to remove redundancy from the quantizer output symbols and to map them on to a bit stream. In the source decoder an 'inverse' block called a decoding block maps the received bit stream on to quantizer output symbols. Coding and decoding blocks are discussed in Chapter 3.

Bit-rate control blocks

The bit stream from a coding block may have a variable rate. Many channels only accept fixed bit rates, so bit-rate control may be required. Methods of bit-rate control will be further introduced in Chapter 6, which discusses adaptive source coding.

In the discussion in Chapters 3–5 it is assumed that the signal has unchanging statistical properties or, in other words, that it is a stationary stochastic signal [10]. Usually this is not the case. The properties of signals may vary in time. For instance, changes in power, frequency distribution and in the fundamental frequency can be observed in speech signals. Since source-coding systems work better if they are tuned to certain properties, they are often made adaptive to the signal. Aspects of adaptive source coding are discussed in Chapter 6.

---

[1]Inverse is between quotation marks because it may not be a strictly mathematical inverse.

1.8 Problems

Problem 1 Assume that the quantizer of Example 2 maps its input on to its output according to the following table:

input	output
≤ -5	-6
-4,-3	-4
-2,-1	-2
0	0
1,2	2
3,4	4
≥ 5	6

The $x[i]$, $y[i]$, $\hat{y}[i]$ and $\hat{x}[i]$ are the input of the source encoder, the output of the source encoder, the input of the source decoder and the output of the source decoder, respectively. At instant $i = 0$, they are given by

$$x[i] = y[i] = \hat{y}[i] = \hat{x}[i] = 0.$$

Complete the following table:

n	$x[i]$	$\hat{y}[i]$	$\hat{x}[i]$
0	0	0	0
1	5		
2	6		
3	4		
4	6		
5	8		
6	8		
7	10		
8	11		

What can you say about the distortion of this codec?

Problem 2 Assume that the quantizer of Example 2 has the same characteristic as in Problem 1. The $x[i]$, $y[i]$, $\hat{y}[i]$ and $\hat{x}[i]$ are also defined in Problem 1. At instant $i = 0$, they are given by

$$x[i] = y[i] = \hat{y}[i] = \hat{x}[i] = 0.$$

Complete the following table:

18 Source coding

n	$x[i]$	$\hat{y}[i]$	$\hat{x}[i]$
0	0	0	0
1	20		
2	20		
3	20		
4	20		
5	21		
6	20		
7	20		
8	20		

Now, what can you say about the distortion of this codec? Looking at the topics of Section 1.6, what can you say about this codec?

Problem 3 Assume that the quantizer of Example 2 has the same characteristic as in Problem 1. The $x[i]$, $y[i]$, $\hat{y}[i]$ and $\hat{x}[i]$ are also defined in Problem 1. At instant $i = 0$ they are given by

$$x[i] = y[i] = \hat{y}[i] = \hat{x}[i] = 0.$$

Due to an uncorrected channel error at instant $i = 5$, a wrong value for $\hat{y}[i]$ is received. Complete the following table:

n	$x[i]$	$y[i]$	$\hat{y}[i]$	$\hat{x}[i]$
0	0	0	0	0
1	20			
2	20			
3	20			
4	20			
5	21		-6	
6	20			
7	20			
8	20			

Looking at the topics of Section 1.6, what can you say about this codec?

Problem 4 In Example 4 three types of data are transmitted: the divider, the value of the first pixel on a line, and the pixel differences. Analyse the influence of channel errors on each of them. Assume that the effect of a channel error is inversion of one bit.

2

The basic blocks of a coding system

2.1 Introduction

The basic blocks of a waveform coding system, as these were referred to in Section 1.7, can be identified in all waveform coding systems. Each of these blocks plays an important role in connection with removing redundancy, introducing distortion, exploiting irrelevancy and bit-rate control.

There are two groups of basic blocks in a source-coding system. The first consists of transformation, quantization and coding blocks in the source encoder, and inverse-transformation, reconstruction and decoding blocks in the source decoder. The second group, which is only present if a fixed bit rate is desired, consists of blocks for bit-rate control.

In Section 2.2 transformation, quantization and coding blocks, and their counterparts in the source decoder, are further introduced. Bit-rate control methods will not be further introduced in this chapter but in Chapter 6, which is about adaptivity in source-coding systems.

2.2 Coding, quantization and transformation

2.2.1 An example

Coding, transformation and quantization blocks in the source encoder, and their counterparts in the source decoder, are introduced by means of the following, somewhat extensive, example of a simple source-coding technique called block companding. It is also known as block-companded coding or adaptive pulse-code modulation (APCM) [4, page 192]. The term APCM is used here. APCM is important because it often occurs as a part of more complicated source-coding systems.

Example 5 The input signal of the APCM encoder is real-value. It is divided into blocks of N samples each. Since these blocks are treated separately, it will suffice to consider only one block. The samples of this block are denoted by $x[i]$, $i = 0, \ldots, N-1$.

As a first operation the samples of the block are stored in a buffer. The maximum absolute value

$$|x|_{\max} = \max_{i=0,\ldots,N-1} |x[i]|,$$

also called peak value, is computed. The $x[i]$ are divided by $|x|_{\max}$, resulting in a block of scaled samples

$$y[i] = \frac{x[i]}{|x|_{\max}}, \; i = 0, \ldots, N-1,$$

with, of course,

$$-1 \leq y[i] \leq 1.$$

The idea is to transmit the $|x|_{\max}$ and the $y[i]$ and to reconstruct the input signal in the source decoder. The advantage of this procedure is explained later in this example. Division into blocks and subsequent scaling can be seen as a transformation. Inverse scaling and conversion of the blocks into a sequence of samples in the source decoder is then the inverse transformation.

Since the $|x|_{\max}$ and the $y[i]$ are real numbers, transmission would require infinitely many bits for each of them. Instead of $|x|_{\max}$ and the $y[i]$, symbols chosen from finite sets are therefore transmitted. The mapping of $|x|_{\max}$ and the $y[i]$ on to these symbols is called quantization. It is done by the quantization block. The symbols which the quantizer-input values are mapped on to are called quantizer-output symbols. A quantizer is denoted by Q. If there are M distinct quantizer-output symbols, then at most $\lceil \log_2(M) \rceil$ bits are needed to code each symbol, where $\lceil \log_2(M) \rceil$ denotes the first integer greater than or equal to $\log_2(M)$. In the source decoder, a reconstructor, denoted by R, maps each quantizer-output symbol on to a number which is an approximation of the quantizer-input value. Since many quantizer-input values are mapped on to the same quantizer-output symbol, there can be no such thing as inverse quantization and errorless reconstruction in the source decoder is impossible.

The quantization block of this example contains N identical quantizers for the scaled samples $y[i]$, also called sample quantizers, and one for the peak value $|x|_{\max}$, also called the peak quantizer.

Quantization of the scaled samples $y[i]$ is discussed first. A set of M_y values $\{l_1, \ldots, l_{M_y}\}$ in the interval $[-1, 1]$ has been selected. To each l_j corresponds one quantizer-output symbol a_j, which will be transmitted to the source decoder. The symbol a_j can, for instance, be the index of the quantization level. In this particular example the quantizer-output symbol corresponding to the l_j closest to $y[i]$ is chosen. Since there is a one-to-one correspondence between the l_j and the a_j, the reconstructor in the source decoder can recover the l_j and will assign

$$\hat{y}[i] = l_j.$$

The $l_j, j = 1, \ldots, M_y$ are called quantization levels. There are various ways of choosing them, resulting in various types of quantizers. Here it is assumed that they are uniformly spread over the interval $[-1, 1]$. This type of quantizer is called a uniform quantizer. Different types of quantizers are discussed in Chapter 4. The mapping from $y[i]$ on to

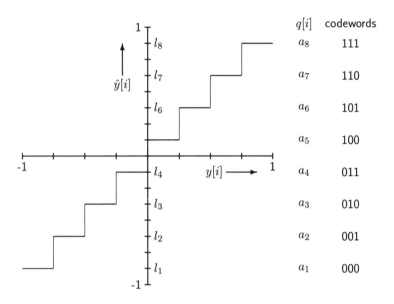

Figure 2.1 *Example of a quantization characteristic.*

$\hat{y}[i]$ is called a quantization characteristic. An example of the quantization characteristic of an eight-level uniform quantizer is shown in Figure 2.1.

A similar procedure is used to quantize $|x|_{\max}$, which is reconstructed as $\widehat{|x|}_{\max}$. However, there are some differences from the quantization of the $y[i]$. Firstly, it is important to use an exact inverse scaling in the source decoder. This can only be achieved if $\widehat{|x|}_{\max}$ is used as a divisor in the source encoder. Secondly, if it is important to have

$$-1 \leq y[i] \leq 1,$$

then $\widehat{|x|}_{\max}$ must be chosen such that

$$\widehat{|x|}_{\max} \geq |x|_{\max}.$$

In order to transmit them, the quantizer-output symbols must be mapped on to sequences of bits, called binary codewords. In this example the mapping is simple. In fact, it is almost trivial. Later in Subsection 2.2.2 less trivial examples will be given. There are M_y distinct symbols for the samples and M_x for the peak value, which is transmitted only once per block of N samples. The symbols for samples and peak value can be coded with $\lceil \log_2(M_y) \rceil$ and $\lceil \log_2(M_x) \rceil$ bits, respectively. The bit rate R then folllows from

$$R = \frac{N \lceil \log_2(M_y) \rceil + \lceil \log_2(M_x) \rceil}{N} \text{ bits/sample.} \tag{2.1}$$

22   The basic blocks of a coding system

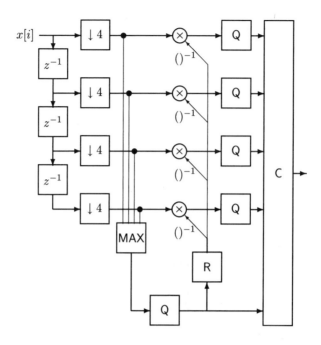

Figure 2.2 *APCM source encoder with a block length of four.*

The block that takes care of mapping quantizer-output symbols on to binary codewords is called the coder, denoted by C. The block that takes care of mapping received binary codewords on to quantizer-output symbols is called the decoder, denoted by D. An example of the mapping of quantizer-output symbols on to binary codewords is shown in Figure 2.1.

Figures 2.2 and 2.3 show block diagrams of an APCM source encoder and decoder, respectively. The block length equals four. More often used and simpler, but less correct, are the source encoder and decoder diagrams of Figures 2.4 and 2.5, respectively. Note that in Figure 2.4 the storage of the samples is absent and that the time-varying behaviour of the source encoder cannot be observed from the diagram. In this figure the blocking of the input signal is done by the block MAX. The peak value can only be estimated correctly if a delay of N samples is introduced in the upper branch just before scaling. This delay is generally omitted. Note also that, instead of five quantizers in Figure 2.2, there are only two in Figure 2.4.

The advantage of APCM over direct quantization of the input samples can be explained as follows. The quantizer introduces quantization errors that result in distortion after source decoding. Because of multiplication by the scaling factor, the error after source decoding is proportional to the signal's peak value. The signal-to-noise ratio (1.5) is therefore more or less constant. This makes this principle suitable for application in music and speech source-coding systems, where a substantially higher noise level can be tolerated in loud passages than in quieter passages.

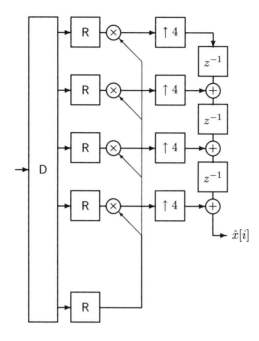

Figure 2.3 *APCM source decoder with a block length of four.*

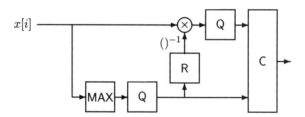

Figure 2.4 *Simplified diagram of APCM encoder.*

Parameters in the system described here are the block length N and the numbers of quantization levels M_y and M_x. The numbers of quantization levels determine the signal-to-noise ratio and the bit rate. The block length N determines the source-coding delay. It also influences the bit rate: it can be seen from (2.1) that the bit rate decreases with increasing N. Finally, N also influences the quality of the output signal, which generally decreases with increasing N. The reason is that for large blocks the peak value is no longer related to the local signal level.
◇

24   The basic blocks of a coding system

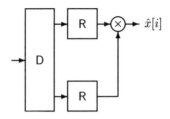

Figure 2.5 *Simplified diagram of APCM decoder.*

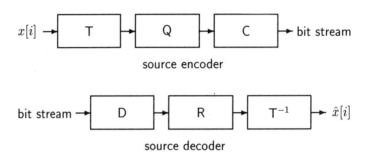

Figure 2.6 *Blocks of a source-coding system.*

Example 5 shows that the source encoder consists of three parts: a transformation block T, a quantization block Q and a coder block C. The source decoder contains their counterparts: a decoder block D, a reconstructor block R and an inverse-transformation block T^{-1}. These blocks are discussed in this book as basic blocks of source-coding systems. Figure 2.6 gives a diagram showing these basic blocks in a source-coding system.

It must be pointed out that Figure 2.6 is not a general diagram. Sometimes the transformation and quantization blocks have memory. This means that their behaviour depends on previous input values and possibly also on previous quantizer-output symbols. If the latter is the case, then there is a feedback from the output of the quantization block to the inputs of the transformation and quantization blocks. This is, for instance, the case in the DPCM coding scheme of Figure 1.3 and in some of the adaptive source systems discussed in Chapter 6. To avoid unnecessary complexity at this moment, this aspect is not discussed further in the present chapter.

A complication of the division into basic blocks is that it is not always unique. In particular, the separation between the transformation and quantization blocks can be put at different places. For instance, in some papers the scaling in an APCM codec is considered to be part of the quantizer, which is then called an adaptive quantizer, whereas in other papers, and in this book, scaling is considered to be part of the transformation.

Blocks T, Q and C have different types of input and output sets. The input of the transformation block consists of samples representing signal values at certain instants. The output values of the transformation block are generally not samples, but just numbers resulting from a mathematical operation. However, they sometimes have a physical interpretation. For instance, it will become clear in Chapters 5 and 12 that a transform coefficient in a transform coder based on a discrete cosine transform can be interpreted as a frequency component of the input signal. The output of a quantizer consists of elements of a finite set. The output of a coder is a sequence of bits.

The functions of these basic blocks are discussed further in Subsections 2.2.2 to 2.2.4.

2.2.2 Coder and decoder blocks

A coder block has two functions. The first is to map quantizer-output symbols on to binary codewords that are transmitted. The simplest way of doing this is to map the quantizer-output symbols on to fixed-length binary codewords, as is done in Example 5. This is often called fixed-length coding. If a quantizer has M distinct quantizer-output symbols this requires codewords comprising $\lceil \log_2(M) \rceil$ bits. The decoder can recover the quantizer output symbols by a table-lookup procedure. As will become clear in the following paragraphs, this is not always the most efficient in terms of bit rate.

The second function of the coder is to decrease the bit rate required to transmit M distinct quantizer-output symbols below $\lceil \log_2(M) \rceil$ bits per symbol. This function is not always present, because it increases complexity and sometimes the gain in bit rate is small.

One approach to decrease the bit rate for the output of a quantizer with M distinct quantizer-output symbols below $\lceil \log_2(M) \rceil$ bits per symbol is illustrated by the following example.

Example 6 Assume in Example 5 that $N = 4$, $M_x = 25$ and $M_y = 5$. If each quantizer-output symbol is mapped on to one binary codeword, the final bit rate is

$$R = \frac{4\lceil \log_2(5) \rceil + \lceil \log_2(25) \rceil}{4} = 4.25 \text{ bits/sample}.$$

Assume now that, before coding, the output symbols of the sample quantizers are formed into two groups of two symbols each. For every group there are $M_y^2 = 25$ possibilities, so that $\lceil \log_2(25) \rceil = 5$ bits are required for each group. This comes down to a bit rate of

$$R = \frac{2\lceil \log_2(25) \rceil + \lceil \log_2(25) \rceil}{4} = 3.75 \text{ bits/sample}.$$

The bit rate can be reduced even further, to 3.5 bits/sample, by forming the quantizer-output symbols into two groups: one consisting of an output symbol from each of three

sample quantizers and one consisting of an output symbol from the fourth sample quantizer and the output symbol from the peak quantizer
◇

The above example shows that, as far as the bit rate is concerned, it is profitable to code groups of quantizer-output symbols together. For an M-level quantizer, coding groups of L samples leads to a bit rate R of

$$R = \frac{\lceil \log_2(M^L) \rceil}{L} = \frac{\lceil L \log_2(M) \rceil}{L} < \frac{L \log_2(M) + 1}{L}. \tag{2.2}$$

For L large enough, R approaches

$$\lim_{L \to \infty} R = \log_2(M). \tag{2.3}$$

This approach complicates coding and decoding. The reader may wonder why so much effort is spent to obtain a gain of at most one bit per sample. The reason is that advanced source coders for audio, video and speech often reduce bit rates to values below one or two bits per sample. In those cases a further gain in bit rate of, for instance, 0.5 bits per sample is quite substantial.

There is another way of reducing the bit rate. The only assumption about the output of a quantizer made so far, is that there are M distinct quantizer-output symbols. Often these values do not have the same probability. If this is the case, and if the probability of each output value is known, a lower bit rate can be achieved. This is illustrated in the following example.

Example 7 Assume in Example 5 that there are eight quantization levels $l_1, \ldots, l_8$ and that the corresponding quantizer-output symbols are $a_1, \ldots, a_8$ with probabilities

$$\frac{1}{16}, \frac{1}{16}, \frac{1}{8}, \frac{1}{4}, \frac{1}{4}, \frac{1}{8}, \frac{1}{16}, \frac{1}{16},$$

respectively. In a long sequence symbols a_4 and a_5 will occur on average twice as often as symbols a_3 and a_6, which will occur twice as often as symbols a_1, a_2, a_7 and a_8. By assigning shorter codewords to the symbols that occur more often, a bit rate can be obtained that is on average lower than $\lceil \log_2(M) \rceil$, where M is the number of distinct quantizer-output symbols. For instance, if the codewords 1111, 1110, 110, 10, 00, 010, 0110 and 0111 are assigned to the symbols $a_1, \ldots, a_8$, respectively, an average bit rate of 2.75 bits/sample is obtained.
◇

Since in Example 7 variable-length codewords are assigned to quantizer-output symbols, this type of coding is called variable-length coding. In this case the coding method is called Huffman coding. Chapter 3 explains Huffman coding in more detail.

It has been shown that coding blocks use two principles to reduce the bit rate. The first is that sequences can be coded with fewer bits per sample than individual samples. The second is that assigning shorter codewords to quantizer-output symbols with a high probability compared to those with a low probability results in a lower bit rate. Combined use of these principles leads to even lower bit rates, but at the expense of greater complexity.

Note that the coders in Examples 6 and 7 reduce the bit rate obtained in Example 5 from

$$R = \frac{N\lceil \log_2(M_y)\rceil + \lceil \log_2(M_x)\rceil}{N}$$

to lower values without introducing errors. If such an errorless reduction is possible, the representation of the quantizer-output symbols is called redundant. Coding blocks can remove this redundancy. The amount of redundancy that is present in the output of the quantizer depends on the quantizer and on the statistics of the quantizer input.

2.2.3 Quantization and reconstruction blocks

In most waveform coders the quantization block is responsible for a substantial part of the reduction of the bit rate. In Example 5 the bit-rate reduction is entirely due to quantization. The quantization block consists of one or more quantizers. Each quantizer maps an output of the transformation block on to a quantizer-output symbol in the set $\{a_1, \ldots, a_M\}$. The number of distinct quantizer-output symbols is (much) smaller than the number of possible quantizer-input values. The quantizer-output symbols can therefore be coded with a smaller number of bits. Often the representation of the input values of a quantizer is so accurate that they can be assumed to be real numbers. This assumption sometimes simplifies the analysis of a quantizer.

In the source decoder the reconstruction block converts received quantizer-output symbols to quantization levels in the set $\{l_1, \ldots, l_M\}$. The mapping of quantizer-output symbols on to quantization levels is a one-to-one mapping.

It has already been observed that many quantizer-input values are mapped on to the same quantizer-output symbol. There is therefore no inverse mapping and errorless reconstruction in the source decoder is impossible. Accuracy of the original representation is lost. Sometimes this accuracy is not required or appreciated by the observer. Part of the representation is then considered irrelevant. Quantizers remove this irrelevancy from the signal's representation.

The quantizer in Figure 1.3 differs from the quantizers Q in Figures 2.2 and 2.4. The quantizer in Figure 1.3 puts out quantization levels whereas the quantizers Q in Figures 2.2 and 2.4 put out quantizer-output symbols. Although it is often found in standard literature, the use of a quantizer such as that in Figure 1.3 is conceptually less correct. Its use can be convenient if only the distortion is of importance and coding of quantizer-output symbols is not considered.

So far, only scalar quantizers, of which the input is a scalar, have been discussed. There is another type of quantizer, called a vector quantizer [7, 21], the input of which is a finite sequence of numbers, also called a vector. Each input vector is mapped on to one quantizer-output symbol. The reconstructor of a vector quantizer converts the quantizer-output symbol to a vector that approximates the original input vector.

At the same bit rate a vector quantizer can generally achieve a lower distortion than a scalar quantizer. However, both the design of a vector quantizer and the quantization procedure are usually much more complicated than in the case of scalar quantization. Vector quantizers are discussed further in Chapter 4.

The function that maps the input of the quantizer on to the set of quantization levels $\{l_1, \ldots, l_M\}$ that are the possible outputs of the reconstructor is called the quantization characteristic. An example is shown in Figure 2.1. The mapping can be described as follows. The range of the quantizer-input values is divided into adjacent intervals. The boundaries between the intervals are called decision levels or decision thresholds, denoted by t_i, $i = 1, \ldots, M - 1$. Furthermore, t_0 and t_M denote the lower and upper bounds of the range of the quantizer-input value, respectively. The intervals are then given by

$$(t_0, t_1], (t_1, t_2], \ldots, (t_{M-2}, t_{M-1}], (t_{M-1}, t_M].$$

Each quantization level corresponds to an interval. A quantizer-input value in the interval $(t_{j-1}, t_j]$ is mapped on to the corresponding quantization level l_j.

Designing a quantizer and a reconstructor consists in choosing the quantization and the decision levels. For a well-designed quantizer-reconstructor pair the distortion after source decoding is acceptable or even imperceptible. Such a design is generally difficult because of a lack of good, perceptually relevant, distortion measures. If the mean-square error (1.4) can be accepted as a distortion measure, then there are methods to design quantizers which, given a certain number of quantization levels and given the probability density function of the input sample $y[i]$, introduce the minimum distortion. This will be further discussed in Chapter 4.

2.2.4 Transformation blocks

There are two important reasons for applying transformations in source coding. The first is that the inverse transformation can transform the quantization errors in such a way that the perceptual distortion is small. This is sometimes called noise-shaping. The second reason is that the use of transformations leads to signal representations that are easy to code after quantization. A further clarification of the second reason is rather mathematical and is deferred to Chapters 3 and 5. Note that the first reason is related to removing irrelevancy and the second to removing redundancy. The first reason for applying transformations is illustrated by Example 5, which is continued below.

Example 5, continued The quantization levels in this example are uniformly spread over $[-1, 1]$. Blocks of samples are adapted to the range of the quantizer by dividing them by their peak values.

The quantization errors in Example 5 are in the range

$$\left(-\frac{1}{M_y}, \frac{1}{M_y}\right].$$

The inverse-transformation block multiplies the quantization errors in each block of N samples by the peak value $\widehat{|x|}_{\max}$. After source decoding, the coding errors are in the range

$$\left(-\frac{\widehat{|x|}_{\max}}{M_y}, \frac{\widehat{|x|}_{\max}}{M_y}\right].$$

A block of output samples is given by

$$\hat{x}[i] = x[i] + e[i], \quad i = 0, \ldots, N-1,$$

where $e[i]$,

$$-\frac{\widehat{|x|}_{\max}}{M_y} < e[i] \leq \frac{\widehat{|x|}_{\max}}{M_y},$$

is the source-coding error. It can be shown that for a large class of signals the signal power in a block is approximately proportional to $\widehat{|x|}_{\max}^2$. It can also be shown that the power of the source-coding error in a block is proportional to

$$\left(\frac{\widehat{|x|}_{\max}}{M_y}\right)^2.$$

Consequently, the signal-to-noise ratio as defined in (1.5) is proportional to M_y^2. Because this is independent of $\widehat{|x|}_{\max}$, the signal-to-noise ratio is constant in time. For music and speech signals, the signal-to-noise ratio determines the perceptual quality. This means that the source-coding errors remain inaudible if M_y is large enough.

Without using the transformation of this example, the power of the source-coding error is constant and the signal-to-noise ratio would be proportional to $\widehat{|x|}_{\max}$. For speech and music the effect would be that in loud passages the signal-to-noise ratio is high, but in quiet passages it is low. A much larger M_y is then required to obtain an acceptable signal-to-noise ratio in soft passages. If the minimal signal-to-noise ratio is used as a distortion measure, then it can be concluded that because of the use of a transformation the bit rate can be reduced at the same distortion level.

◇

The blockwise scaling of Example 5 is a very simple kind of transformation that is often used in combination with other transformation methods. An example of another simple transformation, also often used in combination with other methods, is analog or digital pre-emphasis. The three well-known transformation techniques that are discussed further in Chapter 5 are orthogonal transforms, predictive filtering and subband filtering.

2.2.5 Basic blocks in realizations of source coders

The division into basic blocks is a functional division. The source-coding system is split into parts that can be distinguished on the basis of what they do. This simplifies analysis and design of source-coding systems. In a practical realization these blocks cannot always be easily and unambiguously, identified. This is illustrated by the next example, in which parts of the transformation and quantization and coding are done by one bit shifter.

Example 8 Consider an APCM coder as in Example 5. The input samples are represented by R_s bits. Assume that, instead of $\widehat{|x|}_{\max}$, the value B is computed such that

$$2^{B-2} < |x|_{\max} \leq 2^{B-1}.$$

Note that B is the number of bits needed to represent the $x[i]$. The binary representation of B can be used directly to code the $\hat{x}_{\max}$. Also assume that the quantization levels are K-bit integers in the range $[-2^{K-1}, 2^{K-1} - 1]$. In this case an obvious code for the output symbols of the sample quantizers would consist of the quantization levels themselves: the quantizer becomes a quantizer-reconstructor and the reconstructor in the source decoder can be omitted. Note that $M_y = 2^K$ and that $M_x = \lceil \log_2(R_s - K) \rceil$. A bit shifter scales, quantizes and codes the samples by shifting them to the right over $R_s - B$ bits and then taking only the K most significant bits. The number of bit shifts must be transmitted. Their maximum is $R_s - K$. The source decoder consists of another bit shifter, shifting the received K-bit codewords over $R_s - B$ bits to the left. In this way a bit rate is realized of

$$R = K + \frac{\lceil \log_2(R_s - K) \rceil}{N} \text{ bits/sample.}$$

◇

2.3 Problems

Problem 5 Compute the total delay of the source encoder and decoder of Figures 2.2 and 2.3, assuming that computation of the peak value, scaling and inverse scaling, quantization and reconstruction, and coding and decoding take no time.

Problem 6 Draw a quantization characteristic of the quantizer of Example 8 with $K = 3$. What is the difference compared with Figure 2.1?

Problem 7 Assume in Example 5 that $-X \leq x[i] \leq +X$. In an APCM codec the peak value $|x|_{\max}$ is often mapped on to quantization levels according to

$$\widehat{|x|}_{\max} \in \{\rho^{-M_x+1}X, \ldots, \rho^{-1}X, X\}.$$

Assume $M_x = 5$. Draw quantization characteristics for $\rho = 2$ and $\rho = \sqrt{2}$. Remember that $\widehat{|x|}_{\max} \geq |x|_{\max}$. What is the maximum variation in signal-to-noise ratio in both cases?

Problem 8 Replace the quantizer-reconstructor of Figure 1.3 by a quantizer and a reconstructor in such a way that quantizer-output symbols are transmitted instead of quantization levels.

3

Coding of discrete sources

3.1 Introduction

The tasks of a coding block in a source-coding system are mapping quantizer-output symbols on to binary codewords and reducing the bit rate to one below $\lceil \log_2(M) \rceil$ bits per symbol, where M is the number of distinct quantizer-output symbols. Mapping quantizer-output symbols on to binary codewords without further reducing the bit rate is often performed together with quantization. In that case the quantizer-output symbols are binary codewords of at least $\lceil \log_2(M) \rceil$ bits. This has been demonstrated in Example 8 on page 30. This type of mapping is straightforward and will not be considered here. The topic of this chapter is how coders reduce the bit rate below $\lceil \log_2(M) \rceil$ bits per symbol, which is a more complicated task[1].

In what follows a quantizer is considered to be a discrete source producing a sequence of quantizer-output symbols $q[i]$ with

$$q[i] \in \{a_1, \ldots, a_M\}.$$

The probability of occurrence for every a_j is assumed to be known and denoted by

$$P\{q[i] = a_j\} = p_j,$$

with

$$\sum_{j=1}^{M} p_j = 1.$$

The probability of occurrence of a sequence of N symbols $a_{j_0} \ldots a_{j_{N-1}}$ is denoted by

$$P\{q[i] = a_{j_0}, \ldots, q[i+N-1] = a_{j_{N-1}}\} = p_{j_0 \ldots j_{N-1}}.$$

The $q[i]$ are random, or stochastic, variables, and the sequence $q[i]$ is called a random, or stochastic, process [10]. Note that in the above definitions the $P\{q[i] =$

---

[1]The non-public domain reference [22] has had a great influence on the way this chapter has been written, for which we acknowledge Dirk Kleima. Some of the examples are also taken from this reference.

a_j} are assumed independent of the time index i. If probabilities are independent of time, the stochastic process is called stationary [10]. In a first approach the $q[i]$ are assumed to be statistically independent. This means that knowledge of the value of a certain $q[i]$ does not give any information about the values of the other symbols in the sequence and that

$$p_{j_0 \ldots j_{N-1}} = p_{j_0} p_{j_1} \ldots p_{j_{N-1}}. \tag{3.1}$$

In most practical cases the p_j and the $p_{j_0 \ldots j_{N-1}}$ are unknown. It is shown in Chapter 4 that they depend on the quantization and transformation blocks and on the probability density function of the input signal. When a source-coding system is designed, the transformation block and the quantizer are usually selected first and then the p_j and the $p_{j_0 \ldots j_{N-1}}$ are estimated for a representative class of input signals.

Very important in this chapter is the concept of entropy, which gives the minimum number of bits per symbol that are needed to transmit the $q[i]$. Entropy is informally introduced in Section 3.2.

A coder maps symbols a_j on to binary codewords. Not every set of binary strings can be used as codewords. The set must be what is called uniquely decodable. How this can be achieved is discussed in Section 3.3.

The most important results of the chapter are presented in Section 3.4, where it is shown that for a discrete source producing symbols a_i with probabilities p_i a code with an average codeword length given by the entropy can always be found. Section 3.5 gives methods of constructing codes.

In Sections 3.2 to 3.5 the input symbols of the coder have been assumed statistically independent. In Section 3.6 the more complicated case of coding statistically dependent symbols is discussed.

3.2 Entropy

In this section the entropy of a source producing symbols $a_1, \ldots, a_M$ with probabilities $p_1, \ldots, p_M$ is introduced. The entropy is a very useful measure of the amount of information that a source can transmit. This statement is not proved here, but merely illustrated. A rigid mathematical proof would be long and tedious.

Consider a sequence of N symbols $q[i]$, $i = 0, \ldots, N-1$, with $N \gg 1$. In this sequence a symbol a_j will occur approximately Np_j times. If it is assumed that each symbol a_j occurs *precisely* Np_j times, then the number of possible sequences is given by

$$K = \frac{N!}{(Np_1)! \ldots (Np_M)!}. \tag{3.2}$$

34 Coding of discrete sources

If all K possible sequences have to be coded with binary codewords, then the codewords must consist of B bits, such that

$$2^B = \frac{N!}{(Np_1)!\ldots(Np_M)!}, \qquad (3.3)$$

or

$$B = \log_2\left(\frac{N!}{(Np_1)!\ldots(Np_M)!}\right). \qquad (3.4)$$

By applying Stirling's approximation

$$\log(n!) \simeq n\log(n),$$

we obtain

$$\frac{B}{N} \simeq -\sum_{j=1}^{M} p_j \log_2(p_j) \stackrel{\text{def}}{=} H(q[i]). \qquad (3.5)$$

The expression (3.5) defines the entropy per symbol $H(q[i])$ of the source producing the $q[i]$. For stationary processes the entropy is independent of the time index. Equation (3.5) suggests that the number of bits per symbol required for this source is given by the entropy. The above is not a proof of this statement but merely an illustration. Shannon has given a more fundamental derivation, showing that for the given source there exists a class of 'most probable' sequences that occur with a high probability, the number of which is approximated by $2^{NH(q[i])}$ [23]. To code these 'most probable' sequences $H(q[i])$ bits per symbol are required. The other sequences require more bits, but occur with such a low probability that it does not influence the average bit rate.

In Section 3.4 it is shown that the entropy is indeed a lower bound for the number of bits per symbol needed to code a discrete source.

3.3 Codes

In this section some properties of codes are discussed. A list of M codewords $A_1, \ldots, A_M$, consisting of binary symbols 0 and 1, e.g. the list 1111, 1110, 110, 10, 00, 010, 0110 and 0111 of Example 7 on page 26, is called a binary code. The set $\{0,1\}$ is called an alphabet. In this book only binary codes are considered, therefore binary codes will be referred to henceforth as codes. The number of zeros and ones in a codeword is the codeword length. A concatenation of codewords is called a message. Because codewords consist of zeros and ones, a message is also a sequence of zeros and ones. A code is only useful in a transmission system if every message can be uniquely decoded.

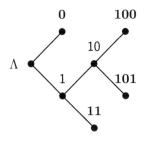

Figure 3.1 *Code tree for* $\{0, 11, 100, 101\}$.

Example 9 Consider the code $\{A_1, A_2, A_3\} = \{0, 01, 10\}$. The message 010 can be decoded as $A_1 A_3$ or as $A_2 A_1$. The code $\{A_1, A_2, A_3\}$ is not uniquely decodable.
◇

A sufficient, but not a necessary, condition for a code to be uniquely decodable is the prefix condition. The prefix condition states that for no two codewords A_i, A_j can a binary sequence S be found such that $A_i S = A_j$. A code satisfying the prefix condition is called a prefix-condition code. The code of Example 9 clearly does not satisfy the prefix condition but the code of Example 7 on page 26 does.

Without proof we state that the codewords of a binary prefix-condition code correspond to the terminal nodes of a binary tree. Figure 3.1 shows this for the code $\{0, 11, 100, 101\}$. The tree can be used for decoding. To begin with we start at the root Λ of the tree. If the first symbol is a zero we follow the upper branch to node 0, otherwise we follow the lower branch to node 1. If, for instance, the first symbol was a one and next the symbol is a zero we proceed from node 1 to node 10, and so on. Finally, we arrive at a terminal node which represents the received codeword. The procedure for the next codeword begins again at the root.

The following three results on uniquely decodable codes, stated without proof, are very important.

Theorem 1 (Kraft, 1949) The codeword lengths $s_1, \ldots, s_M$ of a binary prefix-condition code satisfy

$$\sum_{i=1}^{M} 2^{-s_i} \leq 1. \tag{3.6}$$

Theorem 2 (Kraft, 1949) If the integers $s_1, \ldots, s_M$ satisfy

$$\sum_{i=1}^{M} 2^{-s_i} \leq 1,$$

then there exists a binary prefix-condition code with codeword lengths $s_1, \ldots, s_M$.

The following theorem is stronger than Theorem 1.

Theorem 3 (Macmillan, 1953) The lengths $s_1, \ldots, s_M$ of the codewords of a uniquely decodable binary code satisfy

$$\sum_{i=1}^{M} 2^{-s_i} \leq 1.$$

Inequality (3.6) is generally referred to as Kraft's inequality. According to Theorem 3, uniquely decodable codes satisfy Kraft's inequality. Therefore, because of Theorem 2, there exists a prefix-condition code with the same codeword lengths. Because the codeword lengths are the same, the codes achieve the same bit rate.

The above theorems are used in the next section to derive some of Shannon's source-coding theorems.

3.4 Shannon's source-coding theorems

In this section an expression is derived for the average number of bits per symbol required to code a source signal with given probabilities. Practical coding methods are presented further on in Section 3.5.

In a first approach, codewords are assigned to each of the source symbols $q[i]$. It is shown later that it is useful to assign codewords to groups of source symbols.

Assume that the codeword A_i, with length s_i is assigned to a_i. The probability of a_i is p_i. The average codeword length $\bar{s}$ is given by

$$\bar{s} = \sum_{i=1}^{M} p_i s_i. \qquad (3.7)$$

The lowest average bit rate is achieved with the code that gives the smallest $\bar{s}$ for a given source. It is interesting to derive a lower bound for $\bar{s}$. This lower bound is given by the first half of Shannon's source-coding theorem.

Theorem 4 (Shannon, 1948) The average codeword length $\bar{s}$ of a uniquely decodable binary code satisfies

$$\bar{s} \geq H(q[i]). \qquad (3.8)$$

Proof:

$$\begin{aligned}
\bar{s} - H(q[i]) &= \sum_{i=1}^{M} p_i s_i + \sum_{i=1}^{M} p_i \log_2(p_i) \\
&= \sum_{i=1}^{M} p_i \log_2(2^{s_i} p_i)
\end{aligned}$$

$$\geq \frac{1}{\log(2)} \sum_{i=1}^{M} p_i \left(1 - \frac{1}{2^{s_i} p_i}\right)$$

$$= \frac{1}{\log(2)} \left(1 - \sum_{i=1}^{M} 2^{-s_i}\right) \geq 0 \;\square$$

The inequality $\log(x) \geq 1 - 1/x$ and Theorem 3 have been used in this proof. Equality in (3.8) occurs if

$$2^{s_i} p_i = 1, \; i = 1, \ldots, M.$$

Since the s_i are integers, equality can only be achieved with exceptional p_i, such as those in Example 7 on page 26. Now choose the s_i such that

$$-\log_2(p_i) \leq s_i < -\log_2(p_i) + 1. \tag{3.9}$$

This can be done by choosing $s_i = \lceil -\log_2(p_i) \rceil$. It follows from the left-hand inequality of (3.9) that

$$\sum_{i=1}^{M} 2^{-s_i} \leq \sum_{i=1}^{M} p_i = 1.$$

Then, according to Theorem 2, there must exist a prefix-condition code for this choice of s_i. By multiplying the right-hand inequality of (3.9) by p_i and summing over i we find that $\bar{s} < H(q[i]) + 1$. This is summarized by the second half of Shannon's source-coding theorem.

Theorem 5 (Shannon, 1948) For every set of probabilities $p_1, \ldots, p_M$, there exists a binary prefix-condition code $\{A_1, \ldots, A_M\}$ with codeword lengths $s_1, \ldots, s_M$, which has an average codeword length $\bar{s}$ satisfying

$$\bar{s} < H(q[i]) + 1. \tag{3.10}$$

The implication of Theorems 4 and 5 is that for a given source no code can be found which can be used to transmit data with an average number of bits per symbol less than the entropy, but that certainly a code can be found which can be used to transmit the data with an average number of bits per symbol less than the entropy plus one. This result is valid if one codeword is assigned to every symbol. If groups of symbols are coded together, i.e. if one codeword is assigned to every N symbols, then a code can can be found that is generally closer to the entropy. This is easily seen as follows.

First, a new set of M^N symbols is defined by concatenating N symbols such that

$$a_{j_0 j_1 \ldots j_{N-1}} = a_{j_0} a_{j_1} \ldots a_{j_{N-1}}, \; j_0, j_1, \ldots, j_{N-1} = 1, \ldots, M,$$

with, respectively, probabilities

$$p_{j_0, j_1, \ldots, j_{N-1}} = p_{j_0} \ldots p_{j_{N-1}}. \tag{3.11}$$

The equality (3.11) holds because the symbols are independent. Denote the entropy of the newly defined source by

$$H(q[i]\ldots q[i+N-1]).$$

For $H(q[i]\ldots q[i+N-1])$ we can derive

$$\begin{aligned} H(q[i]\ldots q[i+N-1]) & \\ = -\sum_{j_0,\ldots,j_{N-1}} p_{j_0}\ldots p_{j_{N-1}} & \log_2(p_{j_0}\ldots p_{j_{N-1}}) \\ = NH(q[i]). & \end{aligned} \quad (3.12)$$

If $\bar{s}^{(N)}$ is the average codeword length for a group of N symbols, then, according to Theorem 4,

$$\bar{s}^{(N)} \geq NH(q[i]).$$

Furthermore, according to Theorem 5, there exists a prefix-condition code

$$\{A_1^{(N)},\ldots,A_{M^N}^{(N)}\}$$

with lengths

$$s_1^{(N)},\ldots,s_{M^N}^{(N)}$$

such that

$$\bar{s}^{(N)} \leq NH(q[i]) + 1.$$

The average number of bits per symbol is given by

$$\bar{s} = \frac{\bar{s}^{(N)}}{N}.$$

This leads to the following versions of Shannon's source-coding theorems.

Theorem 6 (Shannon, 1948) The average number of bits per symbol $\bar{s}$ of a uniquely decodable binary code coding N symbols satisfies

$$\bar{s} \geq H(q[i]) \quad (3.13)$$

and

Theorem 7 (Shannon, 1948) For a set of probabilities $p_1,\ldots,p_M$, a prefix-condition code for groups of N symbols exists with an average number of bits per symbol $\bar{s}$, satisfying

$$\bar{s} < H(q[i]) + \frac{1}{N}. \quad (3.14)$$

The last theorem states that the entropy can be approximated arbitrarily closely by coding sufficiently long sequences. In a practical application, however, N cannot be made arbitrarily long, because the complexity of the encoder and decoder and the coding delay will become too large. Some practical coding methods will be described in the next session.

Theorems 4, 5, 6 and 7 illustrate the importance of the concept of entropy for source coding. The entropy is often seen as a measure of information. Intuitively, information can be related to the number of messages that a source can generate. A measure of information could be the number of symbols, in this case bits, that are needed to code these messages. This is the base-two logarithm of the number of messages. Since this number depends on the length of the output sequence of the source, it seems logical to speak of the 'average amount of information per output symbol'. The average number of bits per symbol is the average codeword length $\bar{s}$ of Theorems 4, 5, 6 and 7. These theorems point out that in this sense the information of a source is given by the entropy. If a source is represented with R_s bits per symbol and the source entropy $H(q[i])$ is less than R_s, then the representation is said to be redundant. The redundancy ΔR is given by

$$\Delta R \stackrel{\text{def}}{=} R_s - H(q[i]). \tag{3.15}$$

The above definition formalizes the intuitive concept of redundancy which has been used so far.

3.5 Some coding methods

3.5.1 Huffman coding

The Huffman code [24] is an optimal prefix-condition code. This means that there is no other uniquely decodable code with a smaller average codeword length. A code tree of the kind shown in Figure 3.1 is used in the construction of a Huffman code. The tree for a Huffman code is constructed stepwise by taking together the least probable symbols. Assume that

$$p_1 \geq p_2 \geq \ldots \geq p_M.$$

Now if the code $\{A_1, \ldots, A_{M-1}\}$ is optimal for the set of probabilities

$$p_1, \ldots, p_{M-2}, p_{M-1} + p_M,$$

then it can be shown that the code

$$\{A_1, \ldots, A_{M-1}0, A_{M-1}1\}$$

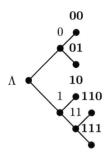

Figure 3.2 *Code tree for the $\{0.30, 0.25, 0.25, 0.10, 0.10\}$. The corresponding codewords are $\{00, 01, 10, 110, 111\}$, respectively.*

is optimal for $p_1, \ldots, p_M$. The Huffman code is now constructed as follows. The probabilities are sorted in order of decreasing magnitude. The codewords corresponding to p_{M-1} and p_M are at two terminal nodes connected to the same node. This node is a terminal node of another code tree for a code with probabilities

$$p_1, \ldots, p_{M-2}, p_{M-1} + p_M.$$

In the next step, therefore, a code tree is constructed for this new code by again sorting the probabilities

$$p_1, \ldots, p_{M-2}, p_{M-1} + p_M$$

in order of decreasing magnitude and two other terminal nodes are found. This procedure is repeated until only two probabilities are left.

Example 10 The following diagram shows how symbols are taken together in every step. The probabilities of the symbols taken together are underlined.

$$
\begin{array}{llllll}
p_1 = 0.30 & 0.30 & \underline{0.30} & & & \\
 & & & 0.55 & & \\
p_2 = 0.25 & 0.25 & \underline{0.25} & & & \\
 & & & & 1.00 & \\
p_3 = 0.25 & \underline{0.25} & & & & \\
 & & 0.45 & \underline{0.45} & & \\
p_4 = \underline{0.10} & & & & & \\
 & 0.20 & & & & \\
p_5 = \underline{0.10} & & & & & \\
\end{array}
$$

The corresponding code tree is shown in Figure 3.2.
◊

The following example shows that it is advantageous to code groups of symbols.

a_i	p_i	$-p_i \log_2(p_i)$	code
a_1	0.2	0.4644	0
a_2	0.8	0.2575	1
		$H(q[i]) = 0.722$	$\bar{s} = 1$

Table 3.1 *Code for a source producing 2 symbols.*

$a_i a_j$	$p_i p_j$	$-p_i p_j \log_2(p_i p_j)$	Huffman code	old code
$a_1 a_1$	0.04	0.1858	111	00
$a_1 a_2$	0.16	0.4230	110	01
$a_2 a_1$	0.16	0.4230	10	10
$a_2 a_2$	0.64	0.4121	0	11
		$H(q[i], q[i+1]) = 1.444$	$\bar{s}^{(2)}/2 = 0.78$	$\bar{s} = 1$

Table 3.2 *Combined code for a source producing 2 symbols.*

Example 11 Consider a source producing symbols a_1 and a_2. The probabilities and the codes are shown in Table 3.1. If groups of two symbols are coded together we obtain the more efficient code that is shown in Table 3.2. In this case coding two symbols together gives a substantial improvement, namely 22%. Taking more than two symbols together is not so useful since not more than another 7% can be gained.
◇

Huffman coding is often used. As has been shown in Section 3.4, it is advantageous to code longer sequences. This is of limited practical value. Huffman coding and decoding are usually done by using lookup tables. For larger sequences of symbols these tables become prohibitively large. Another disadvantage of Huffman coding is that its performance is sensitive to changes in the signal statistics. If the statistics change and the code is not adapted, the bit rate will increase and may even exceed $\lceil \log_2(M) \rceil$ bits per symbol.

It is not necessary for a code to be optimal to satisfy (3.10). This is illustrated by the following example.

Example 12 Consider a source producing symbols a_1, a_2, a_3, with probabilities

$$p_1, p_2, p_3 = 0.05, 0.9, 0.05,$$

respectively. The entropy of this source is given by $H(q[i]) = 0.57$. According to (3.9), codes with codeword lengths

$$s_1 = \lceil -\log_2(p_1) \rceil = 5, \ s_2 = \lceil -\log_2(p_2) \rceil = 1, \ s_3 = \lceil -\log_2(p_3) \rceil = 5,$$

are good enough to satisfy (3.10). In that case the average codeword length $\bar{s}$ is given by $\bar{s} = 1.40$, which is less than $H(q[i]) + 1$. A Huffman code for this source is $\{10, 0, 11\}$.

For this code $\bar{s} = 1.10$.

◇

3.5.2 Runlength coding

Runlength coding is useful if long subsequences, or runs, of the same symbol occur [4, 5]. This is the case, for instance, if the probability-density function of the input of a quantizer shows a sharp peak at zero. Long sequences of zeros can then be expected at the output of the quantizer. That is also the case if binary images, as they occur in fascimile, are coded. The idea of runlength coding is to detect runs of the same symbol and to assign to each run one codeword that indicates its length. An extensive statistical analysis of runlength coding is difficult. Some theoretical results on runlength coding are given in [4]. Properties of this method of coding are illustrated by the following example.

Example 13 The output of a quantizer produces the symbols $\{-1, 0, +1\}$, with probabilities

$$p_-, p_0, p_+ = 0.05, 0.9, 0.05,$$

respectively. The output symbols of the quantizer are statistically independent. The following runs are coded: $+1$, -1, $0+1$, $0-1$, $00+1$, $00-1$, 000. These runs will be denoted by the symbols $A_1, \ldots, A_7$. The first two are length zero runs, the second two are length one runs, etc. The runs occur with probabilities p_+, p_-, $p_0 p_+$, $p_0 p_-$, $p_0 p_0 p_+$, $p_0 p_0 p_-$, $p_0 p_0 p_0$ and are statistically independent. Note that only runs of zeros are considered here, and that the symbol following the run is coded together with the run. This is useful for this particular case, but not generally done in runlength coding. The run

0+10−1000000000+1000000

is coded as

$A_3 A_4 A_7 A_7 A_7 A_1 A_7 A_7.$

A suitable code for the $A_1, \ldots, A_7$ has to be found. Huffman coding is a good candidate here. Note that in this case seven symbols have to be coded in order to code sequences up to length three. Direct Huffman coding of groups of three symbols would lead to no fewer than 27 different codewords. This shows that the coding and decoding complexity of a runlength coder is generally less than the coding and decoding complexity of direct Huffman coding. A runlength code is not necessarily optimal in the sense that it guarantees the smallest possible average codeword length. Example 12, however, has already shown that non-optimal codes may be good enough.

◇

Runlength coding is not only used exclusively for sequences of independent symbols. It can also give good results if the symbols in a sequence are dependent.

3.5.3 Ziv-Lempel coding

Ziv-Lempel coding [5, 25, 26] is a form of universal coding. This means that it adapts to the signal statistics and therefore can be used without measuring statistics and designing codes according to these in advance. It is suitable for sources producing independent symbols as well as for sources that produce dependent symbols. It is often employed in data-compression algorithms used in computers to store data on a disk or diskette.

Assume a source producing M distinct symbols $a_1, \ldots, a_M$. The Ziv-Lempel algorithm codes strings of different lengths with fixed-length binary codewords. Assume also that the codeword length is s. The possible number of input strings is then 2^s. The algorithm sets up string tables in coder and decoder with 2^s entries. The initial table only contains the symbols $a_1, \ldots, a_M$. The procedure is as follows. String $S[k]$ at time instant k is already a member of the string table. Define the current string S by $S = S[k]$.

1. If S contains the last symbol of the sequence, transmit the code of the address in the string table and proceed to step 7.
2. S does not contain the last symbol of the sequence. Append the next symbol $q[k+1]$ to the string and obtain $S[k+1] = Sq[k+1]$.
3. If $S[k+1]$ is in the table, define $S = S[k+1]$ and proceed to step 1.
4. $S[k+1]$ is not in the table. Transmit the code of the string address of S.
5. Add $S[k+1]$ to the string table if there is room left.
6. Define $S = q[k+1]$ and proceed to step 1.
7. Stop transmission.

The decoder always updates its table one step behind the coder. The above is called the Ziv-Lempel-Welch procedure [5]. It is one of the many variations of the Ziv-Lempel algorithm. It is efficient if $2^s \gg M$.

3.5.4 Remarks on the use of coding methods

Alongside Huffman coding, runlength coding and Ziv-Lempel coding, as described above, many variations of coding methods exist. Four possible categories are mentioned here:

Fixed-to-fixed length coding

This is the simplest case in which sequences of source symbols of fixed length are mapped on to binary codewords of fixed length. Fixed-to-fixed length coding is also called fixed-length coding (FLC). A very simple example of fixed-length coding is pulse-code modulation (PCM).

Fixed-to-variable length coding

Huffman coding, in which sequences of source symbols of a fixed length are coded with variable-length codewords is an example of this. Fixed-to-variable length coding is also called variable-length coding (VLC).

Variable-to-variable length coding

Runlength coding, in which sequences of variable lengths of quantizer-output symbols are coded, followed by Huffman coding is an example of variable-to-variable length coding.

Variable-to-fixed length coding

An example of this is Ziv-Lempel coding, in which input sequences of variable length are mapped on to fixed-length codewords.

There are other ways of classifying coding methods, e.g. that used in [27], where some other interesting coding techniques, such as arithmetic coding, are discussed which have not been examined here.

When Huffman coding was discussed, it was observed that the rate may increase if the signal statistics change. To avoid this, coding methods are sometimes made adaptive. Universal coding [28, 29] and Ziv-Lempel coding [25] are examples of adaptive coding methods. Another simple way of adaptive coding consists in letting the coder choose the best out of several Huffman codes. Which code is used must in some way be transmitted to the decoder.

In practical source-coding systems for music, speech and pictures a fixed-to-variable length coding such as Huffman coding is generally used. Runlength coding is also sometimes used. The other methods are less often adopted in these applications, but they are found in, for instance, lossless data compression systems for computer files, or in coding systems for lossless coding of medical images.

The complexity of the coder and decoder largely depends on the number of distinct symbols that have to be coded. This is clearly the case for a fixed-to-variable length code such as the Huffman code. As already stated, coding is done with table-lookup procedures. The number of symbols directly determines the number of table entries. If quantizer-output symbols are coded separately, the number of table entries equals M. If groups of N symbols are coded together, a table with

$$M^N$$

entries is required. Even for small M this number increases rapidly with N.

3.6 Statistically dependent symbols

In this section the coding of signals consisting of statistically dependent sources is considered. The main result is that the entropy per symbol of these sources is less and that, if a good coding method is chosen, a lower bit rate can be achieved than in the case of independent sources with the same probabilities $p_1, \ldots, p_M$ for each symbol.

First, some new concepts and notations must be introduced. As in the preceding section of this chapter a source produces a stationary sequence of symbols

$$q[i] \in \{a_1, \ldots, a_M\}.$$

However, (3.1) is no longer valid since the symbols are statistically dependent. It is useful to define the probability that $q[i] = a_j$, if it is known that $q[i-1] = a_k$. This is the conditional probability [10]

$$P\{q[i] = a_j | q[i-1] = a_k\} = p_{j|k}.$$

Similarly, if it is known that $q[i-1] = a_l$ and that $q[i-2] = a_k$, then

$$P\{q[i] = a_j | q[i-2] = a_k, q[i-1] = a_l\} = p_{j|kl}.$$

Probabilities $p_{x|a\ldots w}$ or $p_{uv|st}$, for instance, can be likewise defined. A result that will often be used is [10]

$$p_{ij} = p_{j|i} p_i. \tag{3.16}$$

Now consider the entropy of a sequence $q[i] \ldots q[i+N-1]$, which is given by

$$\begin{aligned} &H(q[i] \ldots q[i+N-1]) \\ &= -\sum_{j_0, \ldots, j_{N-1}} p_{j_0 \ldots j_{N-1}} \log_2(p_{j_0 \ldots j_{N-1}}) \\ &= -\sum_{j_0, \ldots, j_{N-1}} p_{j_{N-1}|j_0 \ldots j_{N-2}} p_{j_0 \ldots j_{N-2}} \log_2(p_{j_{N-1}|j_0 \ldots j_{N-2}} p_{j_0 \ldots j_{N-2}}) \\ &= -\sum_{j_0, \ldots, j_{N-2}} p_{j_0 \ldots j_{N-2}} \sum_{j_{N-1}} p_{j_{N-1}|j_0 \ldots j_{N-2}} \log_2(p_{j_{N-1}|j_0 \ldots j_{N-2}}) + \\ &\quad -\sum_{j_0, \ldots, j_{N-2}} p_{j_0 \ldots j_{N-2}} \sum_{j_{N-1}} p_{j_{N-1}|j_0 \ldots j_{N-2}} \log_2(p_{j_0 \ldots j_{N-2}}) \\ &= H(q[i] \ldots q[i+N-2]) - \\ &\quad \sum_{j_0, \ldots, j_{N-2}} p_{j_0 \ldots j_{N-2}} \sum_{j_{N-1}} p_{j_{N-1}|j_0 \ldots j_{N-2}} \log_2(p_{j_{N-1}|j_0 \ldots j_{N-2}}). \end{aligned}$$

Define

$$\begin{aligned} &H(q[i+N-1] | q[i] = a_{j_0}, \ldots, q[N-2] = a_{j_{i+N-2}}) \\ &\stackrel{\text{def}}{=} -\sum_{j_{N-1}} p_{j_{N-1}|j_0 \ldots j_{N-2}} \log_2(p_{j_{N-1}|j_0 \ldots j_{N-2}}). \end{aligned}$$

46 Coding of discrete sources

This is the entropy of the most recent output symbol $q[i+N-1]$ if it is known that

$$q[i] \ldots q[i+N-2] = a_{j_0} \ldots a_{j_{N-2}}.$$

In this case, it gives a lower bound to the average number of bits needed to code $q[i+N-1]$. Now define

$$H(q[i+N-1]|q[i]\ldots q[i+N-2])$$
$$\stackrel{\text{def}}{=} \sum_{j_0,\ldots,j_{N-2}} p_{j_0\cdots j_{N-2}} H(q[i+N-1]|q[i]=a_{j_0},\ldots,q[N-2]=a_{j_{i+N-2}}).$$

This is called the conditional entropy of $q[i+N-1]$. It gives a lower bound to the number of bits needed to code $q[i+N-1]$ averaged over all possible $q[i]\ldots q[i+N-2]$. With these definitions we obtain

$$H(q[i]\ldots q[i+N-1])$$
$$= H(q[i]\ldots q[i+N-2]) + H(q[i+N-1]|q[i]\ldots q[i+N-2]).$$

By repeating the above derivation for $H(q[i]\ldots q[i+N-2])$ and by using the fact that the sequence is stationary, we finally obtain

$$H(q[i]\ldots q[i+N-1]) \tag{3.17}$$
$$= H(q[i]) + H(q[i]|q[i-1]) + \ldots + H(q[i]|q[i-N+1]\ldots q[i-1]).$$

It can be shown that

$$H(q[i]) \geq H(q[i]|q[i-1]) \geq \ldots \geq H(q[i]|q[i-N+1]\ldots q[i-1]) \geq 0.$$

Intuitively, this is clear: the number of bits needed to code a symbol will certainly not increase if more is known about previous symbols. From this result and (3.17) it follows that

$$H(q[i]\ldots q[i+N-1]) \leq NH(q[i]), \tag{3.18}$$

or, more generally, for non-stationary sequences,

$$H(q[i]\ldots q[i+N-1]) \leq H(q[i]) + \ldots + H(q[i+N-1]). \tag{3.19}$$

Equality holds in (3.18) and (3.19) if the symbols are independent, which is the case in (3.12). The results (3.18) and (3.19) are important because they show that it is advantageous to make use of the dependency in a discrete source. The entropy of a sequence of N symbols is less than the sum of the entropies of the symbols of the sequence. This implies that a lower bit rate than that given by (3.13) can be achieved coding groups of N symbols.

The question now arises as to how low the bit rates are that can really be achieved for dependent sources. The following derivation gives new bounds for the average codeword lengths that are obtainable for dependent sources.

Since $H(q[i]|q[i-N+1]\ldots q[i-1])$ decreases with increasing N but remains greater than zero, the sequence $H(q[i]|q[i-N+1]\ldots q[i-1])$ converges to a limit for increasing N. Define

$$H(q[i]|q[i-N+1]\ldots q[i-1]) = H_{N-1}(q[i]), \tag{3.20}$$

and

$$\lim_{N\to\infty} H(q[i]|q[i-N+1]\ldots q[i-1]) = H_\infty(q[i]). \tag{3.21}$$

It can be shown that

$$\lim_{N\to\infty} \frac{1}{N} H(q[i]\ldots q[i+N-1]) \tag{3.22}$$
$$= \lim_{N\to\infty} H(q[i]|q[i-N+1]\ldots q[i-1]) = H_\infty(q[i]).$$

If $q[i]$ depends only on $q[i-P]\ldots q[i-1]$, then for all $N \geq P$

$$H_N(q[i]) = H_P(q[i]). \tag{3.23}$$

Without using the assumed independency, (3.13) and (3.14) would read

$$\bar{s} \geq \frac{1}{N} H(q[i]\ldots q[i+N-1]),$$

and

$$\bar{s} < \frac{1}{N} H(q[i]\ldots q[i+N-1]) + \frac{1}{N},$$

respectively. Substitution of (3.22) in these inequalities leads to the following two coding theorems for long sequences of dependent symbols.

Theorem 8 (Shannon, 1948) *The average number of bits per symbol $\bar{s}$ of a uniquely decodable binary code coding a sufficiently long sequence symbols satisfies*

$$\bar{s} \geq H_\infty(q[i]) \tag{3.24}$$

and

Theorem 9 (Shannon, 1948) *For a stationary source a prefix-condition code for groups of N symbols exists with an average number of bits per symbol $\bar{s}$, satisfying*

$$\bar{s} < H_\infty(q[i]) + \frac{1}{N}, \tag{3.25}$$

provided that N is sufficiently large.

Since

$$H_\infty(q[i]) \leq H(q[i]),$$

lower bit rates can be obtained if the dependency is exploited. The redundancy of a sequence of dependent symbols is given by

$$\Delta R = R_s - H_\infty(q[i]). \qquad (3.26)$$

The main conclusion of this section is that lower bit rates can be obtained if the coder exploits the dependency of a source. The problem is how to do this. Coding dependent symbols is generally more difficult than coding independent symbols. Theoretically, the most efficient possibility is to code each new symbol with a Huffman code that depends on the previous symbols. In that case all conditional probabilities must be known in advance. For practical reasons some restrictions have to be placed on these conditional probabilities. Very often it is assumed that each symbol depends only on one or two previous symbols. In that case the process generating the symbols is called a first-order or second-order Markov process, respectively. A more straightforward way is to generate a Huffman code for a group of N symbols, hoping that (3.22) converges fast enough. As has been remarked in Subsection 3.5.4, N cannot become too large since the Huffman table would then become prohibitively large. If $p_{j|j}$ is close to unity for one or more of the a_j, runlength coding maybe a good alternative.

In source-coding systems for physical signals such as speech, music and images large tables are sometimes avoided by making the symbols independent before quantizing and coding. This is done with transformation blocks, which are discussed in Chapter 5.

3.7 Problems

Problem 9 Consider the quantizer of Example 7 as a source. What is the entropy? Does this surprise you?

Problem 10 Reconsider Example 6. What is the entropy that is approached if more and more quantizer-output symbols are coded together? What are the assumed probabilities $p_1, \ldots, p_5$?

Problem 11 Give the derivation of (3.12).

Problem 12 Construct a Huffman code for a source with probabilities

$0.05, 0.05, 0.15, 0.20, 0.25, 0.30$.

Problem 13 Assume that the symbols that follow the runs in Example 13 are coded separately, which means that the symbols $+1, -1, 0, 00, 000$ have to be coded. Are these symbols statistically independent?

Problem 14 A source produces symbols A and B. Assume that a Ziv-Lempel code is used with a string length $s = 3$. Code the sequence

$AABBABBAAABBABBAABBB.$

Play coder and decoder: construct encoder and decoder tables while coding and transmitting. Do the same for $s = 4$. How many bits have you actually used in both cases? If you think it helps, write a computer program!

Problem 15 A stationary source produces symbols 0 and 1. Each transmitted symbol depends only on the preceding one. The dependencies are given by $p_{0|0} = 0.9$, $p_{1|0} = 0.1$, $p_{0|1} = 0.1$ and $p_{1|1} = 0.9$. Furthermore, $p_0 = p_1 = 0.5$. Prove this by solving

$$\begin{pmatrix} P\{q[i] = 0\} \\ P\{q[i] = 1\} \end{pmatrix} = \begin{pmatrix} p_{0|0} & p_{0|1} \\ p_{1|0} & p_{1|1} \end{pmatrix} \begin{pmatrix} P\{q[i-1] = 0\} \\ P\{q[i-1] = 1\} \end{pmatrix}$$

Note that, since the source is stationary, $P\{q[i] = 0\} = P\{q[i-1] = 0\}$ and that $P\{q[i] = 1\} = P\{q[i-1] = 1\}$.

Problem 16 Compute $H_\infty(q[i])$ for the source of Problem 15. What kind of coding algorithm would you use for this source? Why?

4
Quantization

4.1 Introduction

A quantizer, denoted by Q, maps its input values $y[i]$ on to a sequence of quantizer-output symbols, denoted by $q[i]$, which are elements of a set

$$\{a_1, \ldots, a_M\}.$$

In the source decoder the reconstructor R maps these quantizer-output symbols on to a set of quantization levels

$$\{l_1, \ldots, l_M\}.$$

This results in reconstructor-output values, which are denoted by $\hat{y}[i]$. Even though in practical applications the $y[i]$ are represented digitally, their representation is generally so accurate that it can be assumed that they are real. This simplifies the analysis of quantizers. The mapping from the input of a quantizer to the output of a reconstructor introduces errors that cannot be removed. These errors result in distortion at the source decoder's output.

Quantization has two aspects. The first is that a quantizer is a discrete source from which redundancy can be removed by the coding methods described in Chapter 3. This results in a reduced bit rate. The second is that the combination of quantizer and reconstructor introduces coding errors which result in distortion. The two aspects of quantization - the quantizer as a discrete source and the quantizer as an error source - are discussed separately in Sections 4.2 and 4.3, respectively.

There are many types of quantizers and there are many ways of classifying them. In this book no attempt is made to fully classify all types. Some important aspects are illustrated in Section 4.4. An extensive discussion on quantization, including many examples, is given in [4].

We can distinguish quantizers with and without memory. In quantizers without memory the mapping from each input value on to a quantizer-output symbol is independent of previous input values and quantizer-output symbols. In a quantizer with memory this mapping can depend on previous input values and quantizer-output symbols. In this chapter it is mainly memoryless quantizers that are discussed; only the noise-shaping quantizer of Subsection 4.4.4 is an example of a quantizer with memory.

Source coding of a discrete source is done by a coder as has been described in Chapter 3. Source coding of a discrete-time continuous-value source can be done with a combination of a quantizer and a coder. The bit rate and the distortion of such a simple coding system depend on the probability density function of the quantizer-input values and on the design of quantizer and reconstructor. In Section 4.5 a theoretical analysis is given of the bit rate and distortion that, under certain assumptions, can be obtained with a quantizer-coder combination. Section 4.6 relates these results to the rate-distortion theory for Gaussian signals. It is found that a simple coding system, consisting of a quantizer and coder, cannot approach the rate-distortion bound arbitrarily closely. This can only be achieved by means of another type of quantizer called the vector quantizer.

The input of a vector quantizer is a vector, whereas the input of the quantizer discussed so far, the scalar quantizer, is a scalar. The vector quantizer maps its input vector on to a quantizer-output symbol, usually referred to as the index. The vector reconstructor maps the quantizer-output symbol on to a vector. The vector quantizer is discussed in Section 4.7.

Section 4.8 summarizes some of the results of this chapter in order to give some criteria by which to select quantizers for a specific application.

4.2 The quantizer as a discrete source

Mapping on to quantizer-output symbols can be described as follows. The range of the input values is divided into M adjacent intervals. The $M-1$ boundaries separating these intervals are called decision levels or decision thresholds. They are denoted by t_i, $i = 1, \ldots, M-1$. It is convenient to allow $M+1$ decision thresholds and to define t_0 as the lower bound of the range of quantizer-input values and t_M as the upper bound of the range of quantizer-input values. An infinite input range can be expressed by assigning $t_0 = -\infty$ and $t_M = +\infty$. The intervals are given by

$$(t_0, t_1], (t_1, t_2], \ldots, (t_{M-2}, t_{M-1}], (t_{M-1}, t_M).$$

Each interval is represented by one quantizer-output symbol. If the input $y[i]$ of the quantizer is in the interval $(t_{j-1}, t_j]$, the quantizer generates the output symbol a_j. The reconstructor maps a_j on to l_j. Usually this is a one-to-one mapping, therefore

$$P\{q[i] = a_j\} = P\{\hat{y}[i] = l_j\} = p_j. \tag{4.1}$$

The p_j are computed as follows. The probability density function of $y[i]$ is denoted by $p_y(x)$. It is assumed that $y[i]$ is a stationary stochastic process [10]. For many signals $p_y(x)$ is symmetrical about $x = 0$. This assumption is further used in this chapter unless it is stated otherwise. An exception is the case of digital pictures,

52 Quantization

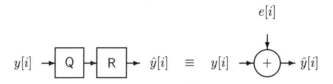

Figure 4.1 *Additive quantization errors.*

where the samples usually attain values in the range $[0, 255]$ and cannot have a probability density function that is symmetrical about zero. The probability of occurrence of the quantizer-output symbol a_j is equal to the probability that the input value is in the interval $(t_{j-1}, t_j]$. Hence,

$$p_j = P\{t_{j-1} < y[i] \le t_j\} = \int_{t_{j-1}}^{t_j} p_y(x) \mathrm{d}x, \ j = 1, \ldots, M. \tag{4.2}$$

The entropy of the quantizer-output symbols $H(q[i])$ is given by

$$\begin{aligned} H(q[i]) &= -\sum_{j=1}^{M} p_j \log_2(p_j) \\ &= -\sum_{j=1}^{M} \left(\int_{t_{j-1}}^{t_j} p_y(x) \mathrm{d}x \right) \log_2 \left(\int_{t_{j-1}}^{t_j} p_y(x) \mathrm{d}x \right). \end{aligned} \tag{4.3}$$

4.3 The quantizer as an error source

The error behaviour of a quantizer is defined by its quantization characteristic Q. This is the function that maps the input $y[i]$ via a set of quantizer-output symbols on to an output value $\hat{y}[i] \in \{l_1, \ldots, l_M\}$. An example has already been shown in Figure 2.1 on page 21.

To analyse quantizer-reconstructor pairs, the output $\hat{y}[i]$ of the reconstructor is sometimes modelled as the input value $y[i]$ of the quantizer plus an additive error $e[i]$, so that

$$\hat{y}[i] = y[i] + e[i]. \tag{4.4}$$

Figure 4.1 illustrates this additive error model. The additive approach is useful when distortion is analysed. In many cases the transformation blocks in source encoder and decoder are each other's inverses, and the transformation block in a source decoder performs a linear operation on the signal $\hat{y}[i]$, resulting in the signal $\hat{x}[j]$. In those cases the output of the source decoder is the sum of the original input signal and another signal, which is the response of the inverse transformation block

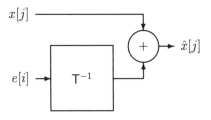

Figure 4.2 *Additive source-decoding model.*

in the source decoder to the quantization error. This approach yields the simple source-decoding model of Figure 4.2. If the quantization noise $e[i]$ can be modelled in some way, then the coding error can be analysed. Examples will be found in Part II. Figure 4.2 illustrates how the inverse transformation block in a source decoder performs noise-shaping. This has already been mentioned in Subsection 2.2.4. Another kind of noise-shaping is discussed in Subsection 4.4.4.

If $y[i]$ is a stochastic process, then the quantization error $e[i]$ is also a stochastic process. Unfortunately, however, its probability density function cannot easily be determined. A special case is discussed in Section 4.4. Expressions can be derived for the mean quantization error $\bar{e}$, which is given by

$$\begin{aligned}\bar{e} &= \sum_{j=1}^{M} l_j p_j - \int_{-\infty}^{+\infty} x p_y(x)\mathrm{d}x \\ &= \bar{l} - \bar{y}, \end{aligned} \qquad (4.5)$$

and for the mean-square error (1.4). When quantizers are considered the mean-square error is often called the quantization-error variance σ_e^2. It is given by

$$\sigma_e^2 = \sum_{j=1}^{M} \int_{t_{j-1}}^{t_j} (l_j - x)^2 p_y(x)\mathrm{d}x. \qquad (4.6)$$

It follows from (4.5) that the quantization error has a zero mean if both the $p_y(x)$ and the l_j are symmetrical about $x = 0$.

Quantization errors can be classified into two types. The first type is due to mapping an input value on to a quantization level when the input value does not exceed the range $[l_1, l_M]$ too greatly. These quantization errors are sometimes referred to as granular noise. The second type of quantization errors occurs when the input values greatly exceed the range $[l_1, l_M]$. These errors are called overload errors, or sometimes overload distortion. If the $y[i]$ have a bounded probability density function, such as a uniform probability density function, overload errors can be avoided by a good choice of the quantizer, or by first scaling the input values. This was the approach of Example 5 on page 19. If the $y[i]$ have an unbounded probability density function, such as a Gaussian probability density function, input

54 Quantization

values can always exceed the range $[l_1, l_M]$. In that case it is convenient to rewrite (4.6) as

$$\sigma_e^2 = \int_{-\infty}^{t_1} (l_1 - x)^2 p_y(x) dx + \int_{t_{M-1}}^{+\infty} (l_M - x)^2 p_y(x) dx + \sum_{j=2}^{M-1} \int_{t_{j-1}}^{t_j} (l_j - x)^2 p_y(x) dx. \tag{4.7}$$

We speak of overload distortion if the errors to input values exceeding $[l_1, l_M]$ become dominant. This is the case if

$$\int_{-\infty}^{t_1} (l_1 - x)^2 p_y(x) dx + \int_{t_{M-1}}^{+\infty} (l_M - x)^2 p_y(x) dx \gg \sum_{j=2}^{M-1} \int_{t_{j-1}}^{t_j} (l_j - x)^2 p_y(x) dx.$$

The following example illustrates the difference between the two types of errors.

Example 14 The input values $y[i]$ of a 7-level and a 63-level quantizer have a zero-mean Gaussian probability density function given by

$$p_y(x) = \frac{1}{\sqrt{2\pi\sigma_y^2}} \exp\left(-\frac{x^2}{2\sigma_y^2}\right), \tag{4.8}$$

where σ_y^2 is the input's variance. The signal-to-noise ratios (1.5)

$$D_{\text{SNR}} = \frac{\sigma_y^2}{\sigma_e^2}, \tag{4.9}$$

of both quantizers are plotted as functions of σ_y^2. The quantization and decision levels have been chosen such that the signal-to-noise ratios of both quantizers are maximal for $\sigma_y^2 = 1$. These quantizers are called pdf-optimized quantizers. They are discussed further in Subsection 4.4.3.

Figure 4.3 shows the signal-to-noise ratios of the quantizers as a function of the input variance σ_y^2. It can be observed that for $\sigma_y^2 < 1$ the signal-to-noise ratio increases with σ_y^2. In this range granular noise errors are dominant. For $\sigma_y^2 > 1$ the signal-to-noise ratio decreases rapidly with σ_y^2 and the overload errors become dominant.
◇

For a large fraction of the area where $\sigma_y^2 < 1$ in Figure 4.3 the signal-to-noise ratio expressed in dBs increases linearly with σ_y^2, also expressed in dBs. More precisely,

$$10 \log_{10}\left(\frac{\sigma_y^2}{\sigma_e^2}\right) = 10 \log_{10}(\sigma_y^2) + \text{constant},$$

from which it follows that in this range the quantization-error variance is almost constant.

For input signals with bounded probability density functions similar curves can be obtained for the signal-to-noise ratio as a function of the input signal's variance. The shape of these curves is roughly the same as in Figure 4.3, only the peaks of

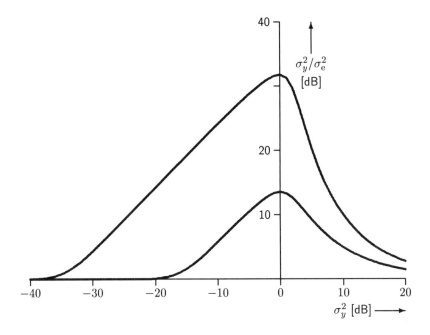

Figure 4.3 *Signal-to-noise ratios of the outputs of a 63-level (upper curve) quantizer and a 7-level quantizer (lower curve) as a function of the input variance for a Gaussian zero-mean input.*

the curves are sharper. The reason is that for signals with a bounded probability density function the region where overload begins is clearly defined, whereas in the unbounded case this region is undefined. Similar curves can also be obtained for other types of quantizers.

The curves of Figure 4.3 show asymmetries about $\sigma_y^2 = 1$. The decrease for $\sigma_y^2 > 1$ is much faster than the increase for $\sigma_y^2 < 1$. This asymmetry is also found for other quantizers and other input signals. It shows that the penalty for overloading a quantizer is greater than the penalty for not fully using its input range. To avoid overload, quantizers should be designed carefully or the input signals should be adapted to the quantizer.

56   Quantization

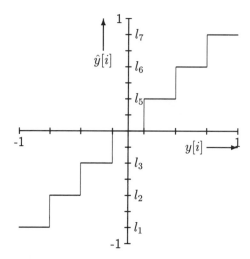

Figure 4.4 *Example of a midtread quantization characteristic.*

4.4 Types of quantizers

4.4.1 Midrise and midtread quantizers

Quantizers are characterized by their quantization and decision levels. The symmetrical quantizer of Example 5 on page 19 has a central decision level $t_4 = 0$. This type of quantizer is called a midrise quantizer. Another type of quantizer is the symmetrical quantizer with a central quantization level $l_{(M+1)/2} = 0$. This type of quantizer is called a midtread quantizer.

The number of levels for a symmetrical midtread quantizer is always odd, whereas the number of levels for a symmetrical midrise quantizer is always even. Figure 4.4 shows an example of a uniform midtread quantizer with seven quantization levels.

The following paragraphs describe some properties of midrise and midtread quantizers.

A general result, presented in [4] and illustrated by the following example, is that the entropy of the quantizer-output symbols is lower for a uniform (see Subsection 4.4.2) midtread quantizer than for a uniform midrise quantizer if the probability density function of the quantizer input is symmetrical about $x = 0$ and has a sharp peak at $x = 0$.

Example 15 The input values $y[i]$ of a uniform quantizer have a Laplacian probability

midrise							
t_1	=	$-t_{13}$	=	-6.00	p_1	=	p_{14}	=	0.0012
t_2	=	$-t_{12}$	=	-5.00	p_2	=	p_{13}	=	0.0025
t_3	=	$-t_{11}$	=	-4.00	p_3	=	p_{12}	=	0.0058
t_4	=	$-t_{10}$	=	-3.00	p_4	=	p_{11}	=	0.0157
t_5	=	$-t_9$	=	-2.00	p_5	=	p_{10}	=	0.0428
t_6	=	$-t_8$	=	-1.00	p_6	=	p_9	=	0.1162
t_7			=	0.00	p_7	=	p_8	=	0.3161
					$H(\hat{y}[i])$			=	2.502
h4f4							
t_1	=	$-t_{14}$	=	-6.50	p_1	=	p_{15}	=	0.0008
t_2	=	$-t_{13}$	=	-5.50	p_2	=	p_{14}	=	0.0012
t_3	=	$-t_{12}$	=	-4.50	p_3	=	p_{13}	=	0.0036
t_4	=	$-t_{11}$	=	-3.50	p_4	=	p_{12}	=	0.0095
t_5	=	$-t_{10}$	=	-2.50	p_5	=	p_{11}	=	0.0259
t_6	=	$-t_9$	=	-1.50	p_6	=	p_{10}	=	0.0706
t_7	=	$-t_8$	=	-0.50	p_7	=	p_9	=	0.1917
					p_8			=	0.3934
					$H(\hat{y}[i])$			=	2.424

Table 4.1 *Decision levels, probabilities and entropies of a midrise and a midtread quantizer.*

density function, which is given by

$$p_y(x) = \frac{e^{-|x|}}{2}. \tag{4.10}$$

The step size of the quantizer is chosen equal to one. Table 4.1 gives a comparison between the decision levels of a 14-level midrise and a 15-level midtread quantizer. For both quantizers the entropies have been computed in accordance with (4.3). It follows that the midtread quantizer has a slightly lower entropy, even though the number of quantization levels is higher. The differences in entropy increase further if the quantization step size increases.
◊

It is shown in [30] that the entropy of a symmetrical midrise quantizer is always greater than one bit if $p(x)$ is symmetric about zero. This makes midrise quantizers useless for very low bit-rate applications.

So far, it has been assumed that the input value of a quantizer is a real number. Sometimes this is not true. It is shown in the next example that in that case a midtread quantizer gives unbiased results, whereas a midrise quantizer does not.

Example 16 Assume that the output of a transformation block is represented with four bits. If a symmetrical probability density function is assumed, then the input values are in the range $[-7, 7]$. A six-level uniform midrise quantizer has decision thresholds $-4.66, -2.33, 0.00, 2.33, 4.66$. A five-level midtread quantizer has decision thresholds

58 Quantization

	midrise		h4f4	
$y[i]$	$\hat{y}[i]$	$e[i]$	$\hat{y}[i]$	$e[i]$
-7	-5.83	1.17	-5.60	1.40
-6	-5.83	0.17	-5.60	0.40
-5	-5.83	-0.83	-5.60	-0.60
-4	-3.50	0.50	-2.80	1.20
-3	-3.50	-0.50	-2.80	0.20
-2	-1.17	0.83	-2.80	-0.80
-1	-1.17	-0.17	0.00	1.00
0	-1.17	-1.17	0.00	0.00
1	1.17	0.17	0.00	-1.00
2	1.17	-0.83	2.80	0.80
3	3.50	0.50	2.80	-0.20
4	3.50	-0.50	2.80	-1.20
5	5.83	0.83	5.60	0.60
6	5.83	-0.17	5.60	-0.40
7	5.83	-1.17	5.60	-1.40

Table 4.2 *Bias of midrise and midtread quantizers with integer inputs.*

$-4.20, -1.40, 1.40, 4.20$. Table 4.2 shows the mapping on to quantization levels for both quantizers. Table 4.2 shows that if a midrise quantizer is used, a bias is introduced in $\hat{y}[i]$ of

$$l_{\frac{M}{2}} P\{y[i] = 0\} = -1.17\ P\{y[i] = 0\},$$

because the input value zero is always mapped on $l_{\frac{M}{2}}$. The bias can be significant, specially if the $y[i]$ have probabilities that are peaked for $y[i] = 0$. Sometimes this problem is solved by randomizing the decision if $y[i] = 0$. If all decisions are randomized, e.g. by adding a small random variable to the $y[i]$, the term dithering is used [4, 31].
◇

Another possible problem with midrise quantizers is due to the fact that there is no level representing zero. If the input values $y[i]$ are very small for some period of time, the output values still attain either the value $l_{\frac{M}{2}}$ or the value $l_{\frac{M}{2}+1}$. In that case the reconstructor produces a sequence of numbers much greater than the input values, which makes the power of the reconstructor output also much greater than the power of the quantizer input. This effect, which can be perceptually disturbing, is illustrated in Figure 4.5.

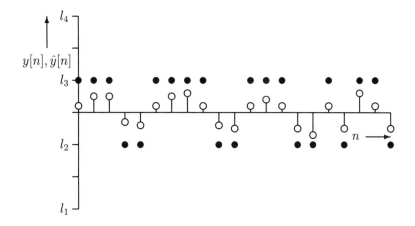

Figure 4.5 *Midrise quantization of small signals, the $y[n]$ being denoted by ○ and the $\hat{y}[n]$ by •.*

4.4.2 Uniform quantizers

A very important class of quantizers is the one called uniform quantizers[1]. Uniform quantizers are so called because the decision and quantization levels are uniformly spread over the input range of the quantizer. In this subsection only symmetrical midtread quantizers, for which M is odd, are discussed. Similar results can be derived for midrise quantizers.

Without loss of generality it is assumed that the input range of the quantizer is $[-1, +1]$. The step size Δ of a uniform quantizer is the distance between two successive quantization levels. It is given by

$$\Delta = \frac{2}{M}.$$

The quantization levels are given by

$$\{l_1, \ldots, l_M\} = \left\{-\frac{M-1}{2}\Delta, \ldots, -\Delta, 0, \Delta, \ldots, \frac{M-1}{2}\Delta\right\},$$

and the decision levels by

$$\{t_1, \ldots, t_{M-1}\} = \left\{(-\frac{M-1}{2} + \frac{1}{2})\Delta, \ldots, -\frac{\Delta}{2}, \frac{\Delta}{2}, \ldots, (\frac{M-1}{2} - \frac{1}{2})\Delta\right\}.$$

---

[1]Uniform quantizers are sometimes wrongly called linear quantizers. Quantization, however, is not a linear operation.

The quantization characteristic is given by

$$\hat{y}[i] = \begin{cases} -\frac{M-1}{2}\Delta, & y[i] \leq -1, \\ \text{round}\left(\frac{y[i]}{\Delta}\right)\Delta, & -1 < y[i] \leq 1, \\ \frac{M-1}{2}\Delta, & y[i] > 1. \end{cases}$$

The uniform quantizer is the only one for which quantization errors can be easily modelled. Note first that, if

$$-1 \leq y[i] \leq 1,$$

the quantization error is bounded as follows:

$$-\frac{\Delta}{2} \leq e[i] < \frac{\Delta}{2}.$$

If the input signal is within the quantizer-input range and if M is large enough, it can be shown for a very large class of signals [4] that the $e[i]$ have as a good approximation a uniform probability density function

$$p_e(x) = \begin{cases} 0, & x < -\frac{\Delta}{2}, \\ \frac{1}{\Delta}, & -\frac{\Delta}{2} \leq x < \frac{\Delta}{2}, \\ 0, & x \geq \frac{\Delta}{2}. \end{cases}$$

The mean quantization error under these conditions is given by

$$\bar{e} = 0,$$

and the quantization-error variance is

$$\sigma_e^2 = \frac{1}{12}\Delta^2. \tag{4.11}$$

Under the same conditions as above it can be shown that the $e[i]$ are, also as a good approximation, statistically independent and independent of the input values $y[i]$. Uniform-quantization noise, therefore, is often modelled as zero-mean, additive white noise with a variance $\frac{1}{12}\Delta^2$.

If the above conditions are not satisfied, the quantization noise becomes correlated with the signal. Subtractive or non-subtractive dithering, explained in [31], can make quantization noise fully uncorrelated.

The fact that uniform quantizers produce quantization noise that is approximately white is one reason for their popularity, since this makes noise-shaping easy, as will be shown in Subsection 4.4.4 and also in Chapter 5. Another reason for the popularity of uniform quantizers is their relatively simple quantization procedure. In non-uniform quantizers very often large tables are used, whereas in the case of a uniform quantizer multiplication by $\frac{1}{\Delta}$ and a rounding operation are sufficient.

4.4.3 Non-uniform quantizers

Non-uniform quantizers have non-uniformly distributed quantization and decision levels. In some situations non-uniform quantizers perform better than uniform ones. Four types of non-uniform quantizers are examined below.

Pdf optimization

Non-uniform quantizers can be optimized in such a way that for a given input probability density function and a given number of levels they give a minimal quantization-error variance. These quantizers are accordingly called pdf-optimized quantizers. The optimization procedure consists in minimizing the quantization-error variance of (4.6) as a function of both the quantization levels $\{l_1, \ldots, l_M\}$ and the decision levels $\{t_1, \ldots, t_{M-1}\}$ for a given number of levels M. If the quantization-error variance is a unimodal function of both the quantization levels and the decision levels, the optimum can be found by setting the partial derivatives of σ_e^2 with respect to $\{t_1, \ldots, t_{M-1}\}$ and to $\{l_1, \ldots, l_M\}$ equal to zero. In that case it follows that

$$t_i = \frac{l_i + l_{i+1}}{2}, \quad i = 1, \ldots, M-1, \tag{4.12}$$

and that

$$l_i = \frac{\int_{t_{i-1}}^{t_i} x p_y(x) \, dx}{\int_{t_{i-1}}^{t_i} p_y(x) \, dx}, \quad i = 1, \ldots, M. \tag{4.13}$$

Quantizers designed in this way are called Max-Lloyd quantizers. The expression (4.12) means that the decision levels are halfway between successive quantization levels and (4.13) means that the quantization level is the centroid of the quantization interval. For many probability density functions $p_y(x)$ (4.12) and (4.13) can be solved iteratively for quantization and decision levels. Algorithms are presented in [32] and [33].

For the same number of levels, pdf-optimized quantizers have a lower quantization-error variance than uniform quantizers. However, some other properties of uniform quantizers are lost. For instance, the quantization error is no longer uncorrelated to the quantizer input, but it is uncorrelated to the quantizer output. As a consequence, it cannot be modelled as additive white noise. Also, the probability density function of the quantization error is no longer approximately uniform.

The result of optimizing a quantizer in this manner is that the density of quantization levels is higher at more probable input levels. For symmetric input probability density functions with a maximum at $x = 0$, this means that more quantization levels are found close to zero. If the number of levels is low, say $M \leq 16$, optimizing is only effective if $p_y(x)$ shows a sharp peak at $x = 0$. This is, for instance, the case for a Laplacian probability density function.

Uniform quantizers can also be pdf-optimized using a similar optimization procedure. They are then called pdf-optimized uniform quantizers. In that case, the step size Δ is chosen in such a way that for a given number of quantization levels the quantization-error variance is minimized.

Entropy optimization

Another way of optimizing quantizers is to minimize the quantization-error variance (4.6) for a given probability density function and a given number of levels at a given entropy of the quantizer-output symbols (4.3). This optimization method is much more complicated than that discussed in the previous paragraphs and will not be explained here. Some results and more references can be found in [4]. Solutions have only been obtained for the mean-square quantization error as a distortion measure [30, 34].

Perceptual optimization

The quantization-error variance is often not directly related to the perceptible distortion. In many cases, therefore, it seems more sensible to optimize the quantizer in such a way that after source decoding perceptible distortion is minimal. This is a difficult problem. Very often the quantization and decision levels have to be adapted by trial and error. Interesting results in the field of image coding are reported in [35].

SNR optimization

If the signal-to-noise ratio of a quantizer is plotted as a function of its input variance, the result is often a diagram such as that given in Figure 4.3. It is characterized by a rather sharp peak. The pdf-optimized quantizers mentioned above also show this behaviour. Another approach to optimizing quantizers is to make the signal-to-noise ratio less sensitive to changes in the input variance. This is achieved by distributing the quantization and decision levels logarithmically over the input range. The bottom-left diagram in Figure 4.10 shows an example of the quantization characteristic of a quantizer of this kind. An example of the signal-to-noise ratio as a function of the input variance for this type of quantizer is shown in Figure 4.6. This figure shows the signal-to-noise ratios as a function of the input variance for a 63-level uniform quantizer and for a 63-level logarithmic quantizer, which is a so-called A-law quantizer [4, 36]. The input signal has a Gaussian probability density function. The uniform quantizer achieves the highest signal-to-noise ratio, but only over a small range. The A-law quantizer yields a lower signal-to-noise ratio, but is almost constant over a wider range. The A-law quantizer is therefore less sensitive to variations in the input level. The μ-law quantizers [4, 37] is another example of a logarithmic quantizer, which has properties that are similar to those of the A-law quantizer.

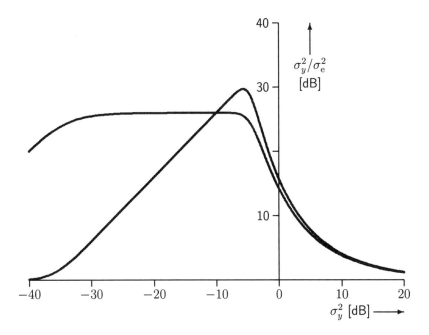

Figure 4.6 *Signal-to-noise ratios of the outputs of a 63-level uniform quantizer (peaked curve) and a 63-level A-law quantizer (flat curve) as a function of the input variance. The input signal has a Gaussian probability density function. The A-law constant is 87.56.*

Logarithmic quantizers are also quite insensitive to variations of the input probability density function. This is illustrated by Figure 4.7, which shows the signal-to-noise ratios of the quantizers of Figure 4.6 with sinusoidal input values

$$y[k] = \sin(k\theta + \psi)$$

with a random phase ψ that is uniformly distributed over $(-\pi, \pi]$. These input values have a probability density function

$$p_y(x) = \begin{cases} \frac{1}{\pi\sqrt{1-x^2}}, & |x| \leq 1, \\ 0, & |x| > 1, \end{cases} \tag{4.14}$$

which is quite different from (4.8). Due to the bounded probability density function, the peak in the signal-to-noise ratio curve of the uniform quantizer has become sharper than in the case of the unbounded probability density function in Figures 4.3 and 4.6. The roughness in the curves is also due to the bounded input probability density function. It can be observed that on average the output signal-to-noise ratio is the same as in Figure 4.6, although the input probability density function is totally different.

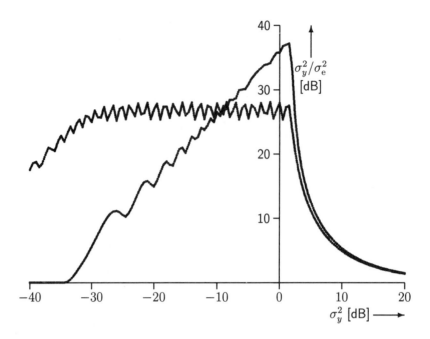

Figure 4.7 *Signal-to-noise ratios of the outputs of a 63-level uniform quantizer (peaked curve) and a 63-level A-law quantizer (flat curve) as a function of the input variance. The input signal is a sinusoid with random phase. The A-law constant is 87.56.*

The direct way to implement a non-uniform quantizer is to determine the interval $(t_{j-1}, t_j]$ to which the input value y belongs, and subsequently to assign the corresponding output symbol a_j to the output. This can, for instance, be done by means of a table. Another approach is to apply first a non-linear invertible function $c(y)$ to the input values, and then to quantize the $c(y)$ uniformly. After reconstruction this results in $\widehat{c(y)}$, from which $\hat{y} = c^{-1}(\widehat{c(y)})$ is computed. For each type of quantizer with a non-decreasing quantization characteristic, a function $c()$ can be found that implements its quantization characteristic. Figure 4.8 shows the basic diagram of such an implementation. This figure shows a resemblance to Figure 2.6 on page 24 if the coder and decoder are omitted from that figure. This shows that the functions $c()$ and $c^{-1}()$ can also be regarded as transformation blocks. Typical examples of functions $c()$ and $c^{-1}()$ are shown, respectively, in the upper left and lower right characteristic of Figure 4.10.

In fact, the previously mentioned A-law and μ-law quantizers are defined in terms of their $c()$ functions. For an A-law quantizer $c()$ is given by

$$c(y) = \begin{cases} y_{\max} \frac{ay/y_{\max}}{1+\ln(a)}, & 0 \leq |y|/y_{\max} \leq 1/a, \\ y_{\max} \frac{1+\ln(a|y|/y_{\max})}{1+\ln(a)} \operatorname{sgn}(y), & 1/a < |y|/y_{\max} \leq 1, \end{cases} \quad (4.15)$$

Types of quantizers

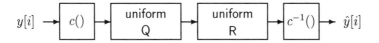

Figure 4.8 *Non-uniform quantizer and reconstructor, realized by means of functions $c()$, $c^{-1}()$ and a uniform quantizer and reconstructor.*

where a is the A-law parameter and $y_{\max}$ denotes the maximum absolute input value. For the μ-law quantizer $c()$ is given by

$$c(y) = y_{\max} \frac{\ln(1 + \mu|y|/y_{\max})}{\ln(1 + \mu)} \operatorname{sgn}(y), \tag{4.16}$$

where μ is the μ-law parameter. For both A-law and μ-law quantizers $c()$ effectively compresses the dynamic range of the input y. It is accordingly called a compression function. Its inverse is called an expansion function. For that reason A-law and μ-law quantizers are also called companding quantizers.

Example 17 Figure 4.9 shows the signal-to-noise ratio of a 15-level A-law quantizer as a function of the input variance. The quantization parameters are $a = 85$, $y_{\max} = 1$. Figure 4.10 shows how this quantizer can be constructed using functions $c()$, $c^{-1}()$ and a 15-level uniform quantizer. The lower left diagram in Figure 4.10 is the desired quantization characteristic. The upper left diagram is the function $c()$. The upper-right diagram in Figure 4.10 shows the quantization characteristic of the 15-level uniform quantizer. Note that the input axis is the vertical axis. The lower-right diagram shows the expansion function $c^{-1}()$. Figure 4.10 shows how the input value $u = 0.3$ is mapped on to $\hat{u} = 0.337$. Direct mapping is via the lower left quantization characteristic; the indirect method uses the upper left compression function, the upper right uniform quantizer and the lower right expansion function.
◇

4.4.4 Noise-shaping quantizers

A quantizer can be combined with a filter which feeds the error back to the quantizer's input. The basic principle is illustrated in Figure 4.11. This type of quantizer is called a noise-shaping quantizer because it changes, or shapes, the power spectral density of the quantization noise. Noise-shaping quantizers combined with dithering are analysed in detail in [38] and [39].

The noise-shaping quantizer is an example of a scalar quantizer with memory, since its output symbol at a certain instant depends on previous quantizer-output symbols.

In the z-domain the noise-shaping filter has a transfer function

$$F(z) = f_1 z^{-1} + \ldots + f_q z^{-q}.$$

66 Quantization

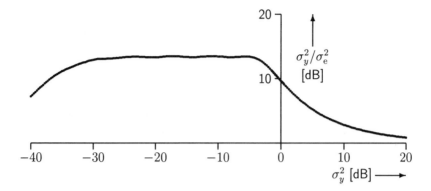

Figure 4.9 *Signal-to-noise ratio of the output of a 15-level A-law quantizer as a function of the input variance. The input signal has a Gaussian probability density function. The A-law constant is 85.*

If we denote the input of the quantizer Q by $y'[i]$, the quantizer adds an error $u[i]$ to $y'[i]$ such that

$$\hat{y}[i] = y'[i] + u[i],$$

or, in the z-domain,

$$\hat{Y}(z) = Y'(z) + U(z).$$

Also,

$$Y'(z) = Y(z) + F(z)U(z),$$

and therefore

$$\hat{Y}(z) = Y(z) + (1 + F(z))U(z).$$

For the z-transform $E(z)$ of the quantization error $e[i]$ of (4.4) we find

$$E(z) = (1 + F(z))U(z). \tag{4.17}$$

If the quantizer is a midtread uniform quantizer and if the quantization levels are chosen such that overload errors cannot occur, then the $u[i]$ can be considered as white noise. In that case the power spectral density of the quantization error is given by

$$S_{ee}(\theta) = |1 + F(e^{j\theta})|^2 \frac{1}{12}\Delta^2 \tag{4.18}$$

where Δ is the quantizer step size. This means that by choosing the $F(z)$ we can shape the power spectral density of the quantization noise in such a way that it is perceptually least disturbing.

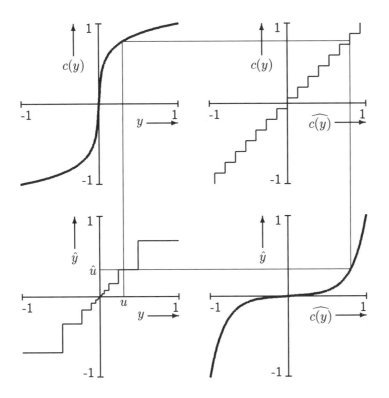

Figure 4.10   *Construction of a 15-level non-uniform A-law companding quantization characteristic. Lower left diagram: A-law quantization characteristic. Upper left diagram: $c()$. Upper right diagram: 15-level midtread quantizer. Lower right diagram: $c^{-1}()$. A-law parameters: $a = 85$, $y_{\max} = 1$. After [4, page 130], adapted by permission of Prentice Hall, Englewood Cliffs, New Jersey.*

It can be shown that

$$\sigma_e^2 = (1 + \sum_k f_k^2)\frac{1}{12}\Delta^2 \geq \frac{1}{12}\Delta^2,$$

which means that noise-shaping increases the quantization-error variance.

Noise-shaping is used in various source-coding systems for music and speech and, with $F(z) = -z^{-1}$, as a part of the digital-to-analog conversion in some compact disc players [40].

Note that noise-shaping is a feedback operation and that the filter and the quantizer must be chosen carefully to avoid instabilities. Quantizer overload can cause such instabilities and, even if it does not, it will mean that the noise-shaper is not working properly.

68 Quantization

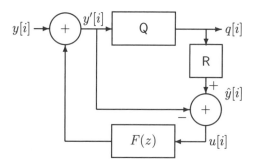

Figure 4.11 *Noise-shaping quantizer.*

4.5 Quantization and coding

This section studies the combination of a quantizer and a coder. More specifically, it relates entropy and distortion to each other. In order to be able to do this we make the following assumptions. The quantizer-input values $y[n]$ are assumed to be statistically independent, to have a zero mean, a variance σ_y^2 and a smooth probability density function. The process $y[n]$ is assumed to be stationary. At a later stage it will also be assumed Gaussian. The quantizer has a uniform step size

$$\Delta \ll \sigma_y. \tag{4.19}$$

The number of quantization levels M is large, which means that

$$M\Delta \gg \sigma_y. \tag{4.20}$$

This justifies the approximation $M = \infty$, which is often used in what follows. The $y[n]$ are reconstructed as $\hat{y}[n] \in \{l_1, \ldots, l_M\}$. Because of assumptions (4.19) and (4.20) the bit rate will turn out to be relatively high. This is why this type of quantization is also referred to as high-rate quantization [7].

The minimum bit rate needed to transmit a quantized signal is determined by the entropy $H(q[i])$ of the quantizer-output symbols; see (4.3). The mean-square error (1.4) is used as the distortion measure. The distortion D_{MSE} is determined by the quantization characteristic and the probability density function of the input values. It is given by (4.6). Because of the one-to-one mapping of quantizer-output symbols on to the quantization levels, (4.1) applies, so that

$$H(\hat{y}[i]) = H(q[i]). \tag{4.21}$$

To avoid the use of too many symbols, only the notation $H(\hat{y}[i])$ will be used.

If the probability density function is sufficiently smooth, then

$$P\{\hat{y}[n] = l_j\} = p_j = \int_{t_{j-1}}^{t_j} p_y(u) du \simeq \Delta \cdot p_y(l_j).$$

On substitution in (4.3) we obtain

$$H(\hat{y}[n]) \tag{4.22}$$

$$= -\sum_{j=1}^{M} p_j \log_2(p_j)$$

$$\simeq -\sum_{j=1}^{M} \Delta \cdot p_y(l_j) \log_2(\Delta \cdot p_y(l_j))$$

$$= -\sum_{j=1}^{M} \Delta \cdot p_y(l_j) \log_2(p_y(l_j)) - \log_2(\Delta) \sum_{j=1}^{M} \Delta \cdot p_y(l_j)$$

$$\simeq -\int_{-\infty}^{+\infty} p_y(u) \log_2(p_y(u)) du - \log_2(\Delta),$$

which is a lower bound to the bit rate for this simple case. The term

$$h(\hat{y}[n]) \stackrel{\text{def}}{=} -\int_{-\infty}^{+\infty} p_y(u) \log_2(p_y(u)) du \tag{4.23}$$

in (4.22) is referred to as the differential entropy. The form of the result (4.22) is attractive, because it shows a resemblance to the definition (3.5) of the entropy of a discrete source. The term $-\log_2(\Delta)$ reflects the influence of the quantization accuracy on the entropy: the more precisely a source is represented, the higher is its entropy.

For a non-uniform quantizer with $\Delta_j = t_j - t_{j-1}$, it can be shown that, under the same conditions of smoothness for the probability density function,

$$H(\hat{y}[n]) \simeq -\int_{-\infty}^{+\infty} p_y(u) \log_2(p_y(u)) du - \mathcal{E}\{\log_2(\Delta_j)\}. \tag{4.24}$$

The operator $\mathcal{E}\{\}$ is the statistical expectation operator. It follows from (4.24) that for non-uniformly quantized signals the entropy depends on the average logarithm of the step size. In [41] it is shown that with assumptions (4.19) and (4.20) the entropy-optimized quantizer is approximately uniform for sufficiently smooth probability density functions.

If the mean-square error is the distortion measure, (4.22) gives the entropy at a distortion of

$$D_{\text{MSE}} = \frac{\Delta^2}{12}.$$

It now follows from the results of Section 3.4 that the quantizer-output symbols $\hat{y}[n]$ can be coded with a bit rate R, expressed in bits per input sample, satisfying

$$h(\hat{y}[n]) - \log_2(\Delta) \leq R < h(\hat{y}[n]) - \log_2(\Delta) + 1. \tag{4.25}$$

If the designer of a source-coding system is ambitious and a higher complexity is not prohibitive, it is possible to obtain

$$h(\hat{y}[n]) - \log_2(\Delta) \leq R < h(\hat{y}[n]) - \log_2(\Delta) + \frac{1}{N} \tag{4.26}$$

70 Quantization

by coding blocks of length N.

As has been remarked in Subsection 3.5.4, a measure for the coding complexity is the number of symbols that have to be coded. If Huffman coding is used, this determines the number of entries in the coding table. Every realistic quantizer has a finite number of levels, which is chosen in such a way that overload distortion is either negligible or acceptable; this is reflected in (4.20). We assume that the number of quantizer-output symbols M satisfies

$$M = K \frac{\sigma_y}{\Delta}. \tag{4.27}$$

Because of (4.19) and (4.20), both K and σ_y/Δ have to be sufficiently high. Combined coding of N quantizer-output symbols requires an encoding table with

$$M^N = K^N \left(\frac{\sigma_y}{\Delta}\right)^N \tag{4.28}$$

entries. This number grows rapidly with increasing N.

The results of this section can be made more precise for Gaussian signals. If quantized samples are coded independently, the entropy $H(\hat{y}[n])$ determines the bit rate. The probability density function of a sample of a zero-mean stationary Gaussian signal $y[n]$ with variance

$$\mathcal{E}\{(y[n])^2\} = \sigma_y^2$$

is given by (4.8). On substitution of (4.8) in (4.22) we obtain

$$\begin{aligned} H(\hat{y}[n]) & \tag{4.29} \\ &\simeq \frac{1}{2}\log_2(2\pi e) + \log_2(\sigma_y) - \log_2(\Delta) \\ &= \frac{1}{2}\log_2(2\pi e) + \log_2(\frac{\sigma_y}{\Delta}) \\ &= 2.05 + \log_2(\frac{\sigma_y}{\Delta}). \end{aligned}$$

For a given variance the Gaussian probability density function (4.8) maximizes the differential entropy (4.23).

4.6 Relation with the rate-distortion theory

The results of Section 4.5 can be related to results from the rate-distortion theory [6, 7]. The rate-distortion theory aims to compute the minimum bit rate needed to code a signal at a given distortion. Only for a small number of combinations of input probability density functions and distortion measures can explicit expressions be obtained. Gaussian probability density functions and the mean-square error D_{MSE}

of (1.4) form such a combination. The derivation is difficult [7], so only results are presented here.

Let the bit rate at a given mean-square error $D_{\text{MSE}} = D$, expressed in bits per input symbol, of a source-coding scheme be denoted by R. Then for each input probability density function $p_y(x)$ there exists a bound, called the rate-distortion bound $R_B(D)$, such that

$$R \geq R_B(D). \tag{4.30}$$

Figure 1.4 on page 13 gives an example of a rate-distortion bound. For the case that the input probability density function is Gaussian with variance σ_y^2 and that the input samples are independent, the rate-distortion bound is given by

$$R_B(D) = \max(0, \frac{1}{2}\log_2(\frac{\sigma_y^2}{D})). \tag{4.31}$$

The rate-distortion bound of the preceding paragraph can be compared with the results (4.29). Note that it follows from

$$D = \frac{\Delta^2}{12}$$

that

$$\log_2(\Delta) = \frac{1}{2}(\log_2(D) + \frac{1}{2}\log_2(12)).$$

On substitution of this result in (4.29) we obtain

$$H(\hat{y}[n]) \simeq R_B(D) + \frac{1}{2}\log_2\left(\frac{e\pi}{6}\right) \simeq R_B(D) + 0.255. \tag{4.32}$$

Hence, for Gaussian inputs, uniform quantizers with a step size that is small compared to σ_y, which have many quantization levels, produce output symbols that can be coded with a bit rate that is 0.255 bit above the rate-distortion bound. The curve $R_B(D) + 0.255$ is called the Gish-Pierce asymptote [30, 41]. In all preceding derivations it has been assumed that an infinite number of quantization levels was available. In the realistic case of a finite number of levels, these results remain valid if overload distortion is negligible. This means that the quantizer-input range must be sufficiently large. This condition can be further relaxed. It is demonstrated in [30] that the curve $R_B(D) + 0.255$ is the performance envelope for all scalar quantizers with Gaussian inputs if $R > 1$. A Gish-Pierce asymptote can also be computed for some other non-Gaussian sources [30].

The rate-distortion bound and the Gish-Pierce asymptote for a Gaussian independent source are shown in Figure 4.12. All scalar quantizers with rates greater than one bit per symbol can be represented by points above the Gish-Pierce asymptote. Examples of quantizers with rates less than one bit per symbol are given in [30].

72 Quantization

Points a and b in Figure 4.12 represent the performance of a 63-level Max-Lloyd quantizer. The distortion is -31.7 dB. The 63 quantizer-output symbols can be mapped directly on to 6 bit symbols. This results in point a. The entropy of the quantizer-output symbols is 5.69. This lower bound to the bit rate can be approximated arbitrarily closely by an appropriate Huffman code, which results in point b.

Points c and d in Figure 4.12 represent the performance of a 63-level pdf-optimized uniform quantizer. The distortion is -29.7 dB. The step size is 0.1055. The 63 quantizer-output symbols can be mapped directly on to 6 bit symbols. This results in point c. The entropy of the quantizer-output symbols is 5.29. This lower bound to the bit rate can be approximated arbitrarily closely by an appropriate Huffman code, which results in point d.

Both points b and d are close to the Gish-Pierce asymptote. The pdf-optimized uniform quantizer (point d) is somewhat closer to it than the Max-Lloyd quantizer (point b). At a given distortion D, the Gish-Pierce asymptote can be approximated arbitrarily closely with a uniform quantizer with step size $\Delta = \sqrt{12D}$ and a number of levels that is large enough to render overload distortion negligible. For the 63-level pdf-optimized quantizer this comes down to $\Delta = 0.1133$.

The rate-distortion bound can be approximated arbitrarily closely for any $D \geq 0$, but only with vector quantizers for vectors of sufficient length [6, 7]. Vector quantization is further explained in Section 4.7.

4.7 Vector quantization

The quantizers explained in Sections 4.2–4.6 quantize each input value individually. They are also called scalar quantizers. It has been explained in Section 4.5 that by variable-length coding of large blocks of scalar quantized samples it is possible to approximate the Gish-Pierce asymptote, which is somewhat above the rate-distortion bound. By applying the appropriate kind of vector quantization it is possible to approximate the rate-distortion bound arbitrarily closely.

The principle of vector quantization is explained in Subsection 4.7.1. A comparison between scalar and vector quantization is made in Subsection 4.7.2. As will become clear in Subsection 4.7.1, a vector quantizer uses a set of vectors called a 'codebook'. Codebook design is briefly discussed in Subsection 4.7.3. Finally, several different types of vector quantizers are discussed in Subsection 4.7.4. For interesting reviews of vector quantization the reader is referred to [21, 42, 43].

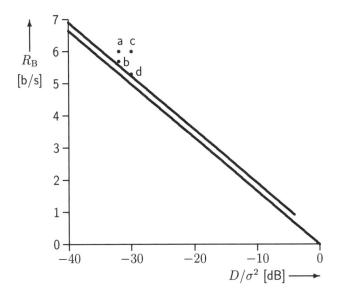

Figure 4.12 Rate-distortion bound (lower curve) and Gish-Pierce asymptote (upper curve) for a Gaussian source. Points a,b,c and d represent different quantizers (a: 63-level Max-Lloyd, 6-bit representation, b: 63-level Max-Lloyd, Huffman-coded, c: 63-level, pdf-optimized uniform, 6-bit representation, d: 63-level pdf-optimized uniform, Huffman-coded).

4.7.1 Principle

Figure 4.13 shows the basic scheme of a vector quantizer. The encoder consists of a codebook containing L codebook vectors l_j, $j = 0, \ldots, L-1$, and of a comparison unit that compares input vectors and codebook vectors. The decoder has a memory containing a copy of the codebook in the encoder.

Quantization is performed as follows. In the encoder N successive input values are grouped into a vector $\mathbf{v}$. This vector is compared with each of the L codebook vectors using a distance measure $d(\mathbf{v}, l_j)$. The codebook vector l_i for which $d(\mathbf{v}, l_i)$ is minimum is chosen. The index $i \in \{0, \ldots, L-1\}$ of the chosen codebook vector is produced at the output of the comparison unit. The index i is the vector quantization equivalent to the quantizer-output symbol of the scalar quantizer.

The index i is coded and transmitted. Fixed-length coding is generally used. In the decoder the quantized vector l_i is reconstructed by a simple table-lookup procedure in the codebook, using the received index i as the lookup address.

The number L of codebook vectors is usually chosen equal to a power of two, i.e. $L = 2^K$, with K a non-negative integer. Assuming fixed-length coding of the

74 Quantization

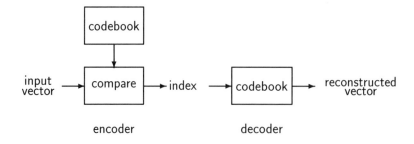

Figure 4.13 *Encoder and decoder of full-search VQ.*

indices with binary codewords, the bit rate R of the vector quantizer is given by

$$R = \frac{\log_2(L)}{N} = \frac{K}{N} \quad \text{bit/element}. \tag{4.33}$$

If, for example, the input vector has 16 elements, and the codebook consists of 256 vectors, the input vector is coded at a rate of 0.5 bit/element. Here we see a first important difference between scalar and vector quantizers. A scalar quantizer followed by fixed-length coding can not operate at bit rates less than 1 bit per input value, whereas a vector quantizer can.

The vector quantizer in Figure 4.13 is often called a full-search vector quantizer. It operates directly on the input vectors and uses a full-search procedure to choose the closest codebook vector, i.e. it compares the input vector with all possible codebook vectors. Other types of vector quantizer are described in Subsection 4.7.4. In what follows, the terms vector quantization and vector quantizer may sometimes be abbreviated as VQ.

Each input vector $\mathbf{v}$ of N elements can be considered as a point in an N-dimensional space. The L codebook vectors $\mathbf{l}_i$, $i = 0, \ldots, L-1$, are also points in this space. For each point $\mathbf{p}$ in the N-dimensional space the closest codebook vector can be calculated. The set of points which are closest to codebook vector $\mathbf{l}_i$ is denoted as C_i. This set is often called a cell. Thus the N-dimensional space is partitioned into L cells C_i, $i = 0, \ldots, L-1$. Cell C_i consists of all points $\mathbf{p}$ that are located closer to $\mathbf{l}_i$ than to any other codebook vector, i.e.

$$\mathbf{p} \in C_i \text{ iff } d(\mathbf{p}, \mathbf{l}_i) \leq d(\mathbf{p}, \mathbf{l}_j) \quad \text{for all } j \neq i. \tag{4.34}$$

Vector quantization comes down to identifying the cell to which the input vector belongs. The index of the cell is coded and transmitted. An example of a partitioning into cells is given in Subsection 4.7.2.

4.7.2 Vector versus scalar quantization

Scalar and vector quantizers differ in a number of aspects. An important difference is the performance in terms of bit rate versus quantization accuracy. It has already been mentioned in Section 4.7.1 that a vector quantizer can approximate the rate-distortion bound arbitrarily closely, whereas a scalar quantizer cannot. The difference in performance is further illustrated by the following example.

Example 18 A stationary vector source generates real-value vectors $\mathbf{v} = (v_1, v_2)^T$. The probability density function $p(\mathbf{v})$ is shown in Figure 4.14. The corresponding probabilities $p(v_1) = p(v_2)$ are shown in Figure 4.15. A comparison is made between scalar quantization of the two vector elements v_1 and v_2 and vector quantization of the vectors $\mathbf{v}$. In both cases it is assumed that the quantizer-output symbols and the codebook indices are coded with fixed-length coding. The mean-square error is the distance measure.

The optimum scalar quantizer for this example is the Max-Lloyd quantizer which has been explained in Subsection 4.4.3. The dashed lines in Figure 4.16 indicate estimates of the optimal quantization levels for this quantizer. Figure 4.17(a) gives the two-dimensional (v_1, v_2) space, in which the VQ codebook vectors are indicated as dots. The codebook vectors have been determined with the codebook design procedure described in Subsection 4.7.3. It is explained in that subsection how codebooks can be designed from a training set of input vectors by an iterative clustering algorithm. Under certain conditions the optimum codebook can be approximated arbitrarily closely if the training set is large enough. The asymmetry in the positions of the codebook vectors in Figure 4.17(a) is due to the fact that only a limited number of training vectors have been used.

For both the scalar and the vector quantizer the bit rate equals 3 bits per vector element. For a better comparison of the scalar and the vector quantizer, Figure 4.17(b) shows all pairs of scalar quantization levels as dots in the (v_1, v_2) space. The partitioning of the two-dimensional space is also indicated in both Figures 4.17(a) and 4.17(b).

Figure 4.17(b) shows the limitation of scalar quantization. It can only adapt to the probability density function of a vector element by placing the quantization levels somewhat more closely together in regions where the probability density function is high. In this example 25% of the pairs of quantization levels will never be used, since they are in a region where the joint probability density function of the vector element is identical to zero. Figure 4.17(a) shows the advantage of VQ. It adapts to the joint probability density function. No codebook vectors are placed in regions where the joint probability density function is zero. Furthermore, the two-dimensional space is not divided into rectangular cells as in the case of scalar quantization. Because of these features VQ provides, at equal bit rates, a higher quantization accuracy than scalar quantization. Correspondingly for equal quantization accuracies VQ requires a lower bit rate than scalar quantization.
◇

The properties demonstrated in the two-dimensional example above are also valid for vectors of greater lengths, i.e. VQ for vectors of N elements can optimally adapt to the joint Nth-order probability density function of the vector by placing only codebook vectors in that region of the N-dimensional space where the input vectors

76 Quantization

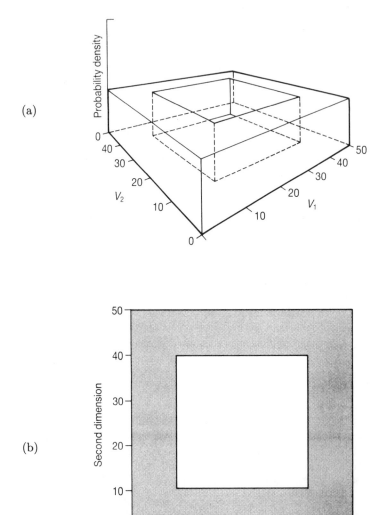

Figure 4.14 *(a) Pseudo-3D view and (b) top view of the two-dimensional probability density function. The shaded area in (b) indicates the region in which the probability density function is constant.*

have a significant probability of occurrence. It can be shown that the rate-distortion bound can be approximated arbitrarily closely by increasing the vector length N

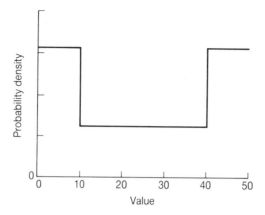

Figure 4.15 *The one-dimensional probability density function.*

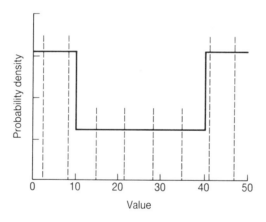

Figure 4.16 *Scalar quantization levels (dashed lines) together with the one-dimensional probability density function.*

[6, 7].

Another important difference between scalar and vector quantization is their complexity. Assume again that the mean-square error is the distortion measure. Scalar quantization is relatively simple. Each input sample is compared with a set of L quantization levels of which the closest level is chosen. This means that the difference between the input sample and the L quantization levels has to be determined, which requires L subtractions. Furthermore, L memory locations are needed to store the quantization levels.

Vector quantization is much more complex. For the storage of a codebook of L vectors of N elements $L \cdot N$ storage locations are needed. From (4.33) it follows

78 Quantization

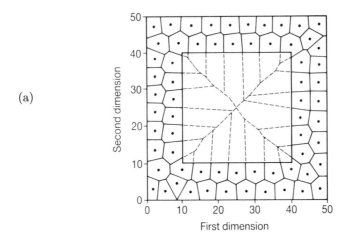

(a)

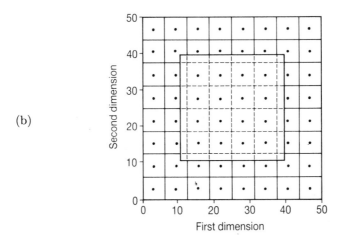

(b)

Figure 4.17 *(a) The set of VQ codebook vectors, and (b) all combinations of scalar quantization levels (dots), plotted in the two-dimensional space. In both figures the lines indicate the partitioning of the space.*

that
$$L = 2^{RN}, \tag{4.35}$$

so that the number of storage locations is equal to

$$N \cdot 2^{RN}. \tag{4.36}$$

This expression shows that the memory requirements increase exponentially with vector length and bit rate.

The computational complexity of vector quantization is also much greater than that of scalar quantization. Using the mean-square error, the number of multiplications and additions needed to quantize one vector element is of the order of

$$L = 2^{RN}. \tag{4.37}$$

This expression shows that computational complexity also increases exponentially with bit rate and vector length.

4.7.3 Codebook design

The codebook vectors have to be chosen such that they are representative of the set of input vectors. Usually, the criterion for the performance of a codebook is the average distortion. Assume that C_i indicates the subset of input vectors quantized to codebook vector l_i. The distortion introduced by quantization of an input vector $\mathbf{v}$ is given by

$$D(\mathbf{v}) = \min_i d(\mathbf{v}, l_i). \tag{4.38}$$

The optimum codebook consists of those vectors l_j, $j = 0, \ldots, L-1$, for which the average distortion $\mathcal{E}\{D(\mathbf{v})\}$, which is defined by

$$\mathcal{E}\{D(\mathbf{v})\} = \int_{\mathbb{R}^N} D(\mathbf{v}) p(\mathbf{v}) d\mathbf{v}, \tag{4.39}$$

is minimum. In this expression $p(\mathbf{v})$ is the probability density function of the input vectors $\mathbf{v}$. Since $p(\mathbf{v})$ is seldom known, direct analytical calculation of the codebook vectors is usually impossible.

In practice the codebook is often designed using an iterative training procedure applied to a large set of input vectors, called a training set. The training procedure is based on the following two necessary conditions for an optimal VQ [21]:

1. Given a certain set of codebook vectors l_j, $j = 0, \ldots, L-1$, the VQ encoder must transmit the index i of the codebook vector l_i that minimizes $d(\mathbf{v}, l_i)$ for each input vector $\mathbf{v}$.
2. The VQ decoder must assign the centroid of the cell C_i to a received index i. This is the vector l_i whose average distance to all vectors in C_i is minimum. In other words it minimizes

$$\int_{C_i} d(\mathbf{v}, l_i) p(\mathbf{v}) d\mathbf{v}.$$

80 Quantization

For arbitrary input vectors and an arbitrary distance measure $d(\mathbf{v},\mathbf{l})$, computation of the centroids can be difficult. For the mean-square error, however, the centroid can easily be calculated. The following paragraphs give a description of a training procedure.

The optimal codebook vectors can be computed from a set of M training vectors $\mathbf{t}_n, n = 1, 2, \ldots, M$ by an iterative clustering algorithm known as the LBG algorithm, which has been named after Linde, Buzo and Gray [44]. Starting with an initial set of codebook vectors, this algorithm iteratively uses the above-mentioned conditions for generating the optimal codebook vectors. The number of training vectors M is typically chosen between 10 and 100 times the codebook size L.

The LBG algorithm is as follows:

1.
- Set iteration counter $m := 0$.
- Choose a set of initial codebook vectors $\mathbf{l}_i(m)$, $i = 0, \ldots, L-1$, and an initial average distortion $D(m)$. These choices will be explained later.

2.
- Set $m := m + 1$.
- Classify the set of training vectors into cells $C_i(m)$ by applying the nearest-neighbour rule, i.e. $\mathbf{t}_n$, $n = 1, \ldots, M$ is classified into $C_i(m)$ if $d(\mathbf{t}_n, \mathbf{l}_i(m)) \leq d(\mathbf{t}_n, \mathbf{l}_j(m))$ for $j = 0, \ldots, L-1$, $j \neq i$.

3.
- Compute the centroids of the training vectors in all cells $C_i(m)$ as the new codebook vectors $\mathbf{l}_i(m)$. Using the mean-square error as the distance measure it can be shown that

$$\mathbf{l}_i(m) = \frac{1}{M_i} \sum_{\mathbf{t}_j \in C_i(m)} \mathbf{t}_j,$$

where M_i is the number of training vectors in $C_i(m)$.

4.
- Calculate the average distortion $D(m)$ between the original and the quantized training set using the codebook vectors $\mathbf{l}_i(m)$, i.e. calculate

$$D(m) = \sum_{i=1}^{L} \frac{M_i}{M} D_i(m),$$

where $D_i(m)$ is the average distortion in cell $C_i(m)$, given by

$$D_i(m) = \frac{1}{M_i} \sum_{\mathbf{t}_j \in C_i(m)} d(\mathbf{t}_j, \mathbf{l}_i(m)).$$

- If the test criterion

$$\frac{D(m-1) - D(m)}{D(m-1)} < \varepsilon$$

is satisfied, i.e. if the relative decrease in distortion is less than a certain given value ε, the procedure is terminated. If not, the procedure is repeated from step 2. For ε, a value of about 0.005 is often used.

After the test criterion of step 4 has been satisfied at a certain value of the iteration counter m, the codebook contains the vectors l_i with $l_i = l_i(m)$. In practice, the algorithm converges within about five to fifteen iterations to a minimum in the average distortion. This minimum, however, can be local and thus the algorithm may result in a suboptimal solution [21, 44].

The initial codebook vectors can be chosen in several ways. A well-known method is the following. The initial codebook vectors used for the design of a codebook with L vectors can be formed by perturbation of the vectors of a codebook with $L/2$ vectors. Each vector of the codebook with $L/2$ vectors is then transformed into two new vectors by adding two small random numbers to it. The distortion obtained with the codebook containing $L/2$ vectors can be taken as the initial distortion for the new codebook. A codebook of length L can be designed starting from a one-vector codebook containing only the centroid of the training set. This vector is perturbed to two new vectors, giving an initial two-vector codebook. From these two vectors, an optimal codebook is designed using the previously described LBG algorithm. The two optimal codebook vectors can again be perturbed, resulting in a non-optimal four-vector codebook, and so on. This procedure is repeated until the required L-vector codebook has been obtained.

4.7.4 Types of VQ

The complexity of VQ is much greater than that of scalar quantization. A considerable amount of memory is required for storage of the codebook. Also, the quantization of a vector requires a large number of calculations. Several methods exist to reduce the complexity. Three of them - multi-stage VQ [45], gain-shape VQ [46] and tree-searched VQ [47] - are discussed in this subsection.

Figure 4.18 gives a schematic representation of the encoder and decoder of a multi-stage VQ with two stages. In the first stage the input vector $\mathbf{v}$ is quantized to $l_{1,i}$ using codebook 1 of length L_1. This codebook can be computed from a training set of input vectors using the LBG algorithm. In stage 2 the residual

$$\mathbf{r} = \mathbf{v} - \mathbf{y}_{1,i}$$

is computed and further quantized to $l_{2,j}$ using codebook 2 of length L_2. Codebook 2 can be computed from a training set of residual vectors with the LBG algorithm. The indices i and j of the codebook vectors are transmitted to the decoder, where a vector

$$\hat{\mathbf{v}} = l_{1,i} + l_{2,j}$$

is constructed.

A VQ with S stages can be constructed by adding additional stages to the above construction. The bit rate R of the S-stage VQ is equal to the sum of the bit rates

82 Quantization

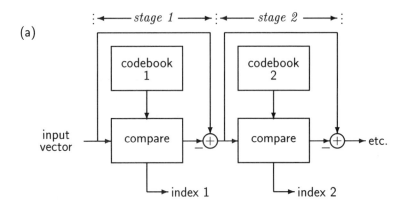

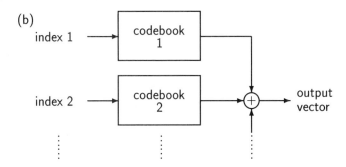

Figure 4.18 *(a) Encoder and (b) decoder of a two-stage cascaded VQ.*

required to transmit the indices of each of the stages. It is given by

$$R = \sum_{i=1}^{S} \frac{\log_2(L_i)}{N} \quad \text{bits/element.} \tag{4.40}$$

At the same bit rate the complexity of the encoder of a multi-stage VQ is much lower than that of the full-search VQ. Assume, for example, that the input vector

v of a two-stage VQ has a length of 16 elements and that codebooks 1 and 2 both contain 256 vectors. The bit rate of the two-stage VQ equals one bit/element, and each input vector has to be compared with 512 codebook vectors. For a full-search VQ with the same bit rate the input vector would have to be compared with no fewer than 65536 codebook vectors. The computational complexity of the decoder of the two-stage VQ is slightly greater than for full-search VQ since two vectors now have to be added.

Another method of reducing the encoding complexity of VQ is gain-shape VQ. Figure 4.19 gives a schematic representation of the encoder and decoder of a gain-shape VQ. First, the gain g of the input vector **v** is calculated as

$$g = \sqrt{\frac{1}{N} \sum_{j=1}^{N} v_j^2}. \tag{4.41}$$

Here it is assumed that the input vector has a zero mean. If this is not the case, the mean can be first subtracted and transmitted separately. This procedure is called mean-separated gain-shape VQ. Subsequently, the gain g is scalar quantized and transmitted as side-information. In the decoder it is reconstructed to $\hat{g}$. Before quantization, the input vector **v** is power-normalized to a vector **n** with elements

$$n_j = \frac{v_j}{\hat{g}}, \ j = 1, \ldots, N, \tag{4.42}$$

The normalized vector **n** is vector-quantized to $\hat{\mathbf{n}} = \mathbf{l}_i$ and the index i is transmitted to the decoder. In the decoder, the $\hat{\mathbf{n}}$ is first retrieved from the codebook and subsequently multiplied by $\hat{g}$.

The VQ described above is called gain-shape VQ since the codebook now contains vectors of which only the shape varies and not the power. All codebook vectors have unit power. Thus for input vectors with different powers but the same shapes, only one codebook vector has to be available. This substantially reduces the number of required codebook vectors. The consequence of this normalization is that the quantization noise depends on the power of the input vector. In fact, the gain-shape VQ has an approximately constant signal-to-noise ratio. Comparison of gain-shape VQ with the APCM scheme of Example 5 on page 19 shows that gain-shape VQ is the vector equivalent of APCM. It is therefore sometimes called vector APCM (VAPCM).

If the number of bits used to code the quantized gain is r_g, the input vector has a length N and the codebook consists of L vectors, the bit rate R of the gain-shape VQ is given by

$$R = \frac{r_g + \log_2(L)}{N} \ \text{bits/element}. \tag{4.43}$$

A third method of vector quantization with reduced complexity is binary tree-search vector quantization. In a binary tree-search VQ, the codebook is organized as a binary tree such as shown in Figure 4.20. The input vector is first compared

84 Quantization

(a)

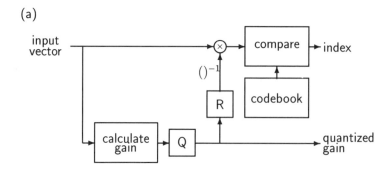

(b)

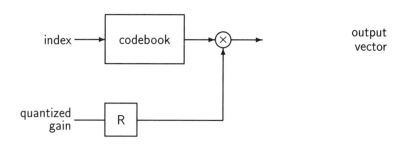

Figure 4.19 *Gain-shape VQ encoder (a) and decoder (b).*

with the two codebook vectors at the top layer of the tree and the closer these is chosen. This choice determines which branch to the second layer is followed. This procedure is repeated until the bottom layer is reached. The index of the vector chosen at the bottom layer is transmitted.

At the same bit rate the codebook of a tree-search VQ has about twice the size of that of a full-search VQ. The number of comparisons required to find the closest codebook vector, however, is significantly lower for the tree-search VQ. Assume, for example, VQ of vectors with a length of 16 elements with a bit rate of 1 bit/element. In full-search VQ, each input vector has to be compared with 65536 codebook vectors. In tree-search VQ, however, not more than 32 comparisons are needed.

The performance of multi-stage, gain-shape and tree-search VQ, in terms of quantization accuracy at a certain bit rate, is generally less than that of full-search VQ. This means that for the same performance multi-stage, gain-shape and tree-search VQ generally require a higher bit rate than full-search VQ.

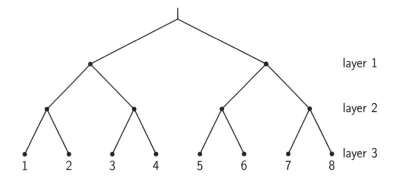

Figure 4.20 *Binary tree-structured VQ codebook.*

4.8 Selecting a quantizer

This section presents some criteria which can be of help when selecting a quantizer. Most criteria are based on the assumption that the mean-square error is the distortion measure and that the input probability density function is given or can be estimated. If this is not the case a satisfactory quantizer, e.g. a perceptually optimal one, can sometimes be determined experimentally.

Midrise quantizers are useless if the required bit rate is less than one bit per sample and if the probability density function of the samples is symmetrical about zero.

Before selecting a quantizer it is important to know whether or not additional variable-length coding can be used. If this is the case, the Gish-Pierce asymptote can be approached by a uniform quantizer and a coder, as is illustrated by point d in Figure 4.12. In many practical applications a uniform quantizer can be implemented more easily than a non-uniform one.

If only fixed-length coding is allowed, a non-uniform pdf-optimized quantizer is a good choice because this type of quantizer gives the smallest mean-square error at a given number of levels. Point b in Figure 4.12 shows that coding the output of pdf-optimized quantizers does not generally lead to approximation of the Gish-Pierce asymptote.

An entropy-optimized quantizer is more complicated to design, but it is capable of approximating the Gish-Pierce asymptote. This is shown in [30]. However, if we are prepared to use more quantization levels, the same result can be achieved with a uniform quantizer. Note that an entropy-optimized quantizer is designed to be always used in combination with a coder.

SNR-optimized quantizers are useful when the signal's variance is not a constant but varies over a certain range. This happens if the signal is not stationary. In practice A-law or μ-law quantizers tuned to the signal are often used. An alternative solution is the block-companded quantizer of Example 5 on page 19.

Vector quantizers are useful since they are capable of approaching the rate-distortion bound. The generation of codebooks and especially the implementation of the quantizer become very complex if codebooks are large and if vectors are long. However, vector quantizers are of increasing importance in speech coding [48].

4.9 Problems

Problem 17 Consider a 7-level uniform quantizer that has an input range $[-1, +1]$. The input signal is $y[n] = \sin(\frac{2\pi}{K}n)$, with K an integer. Choose $K = 6$. Compute the quantization error for a few periods of $y[n]$. How would you characterize this signal? Is it white noise? What have you learned about statements such as 'for a large class...', and what have you learned about choosing test signals?

Problem 18 Look at the quantization errors in Table 4.2 of the midtread quantizer. Are they independent of the input values?

Problem 19 Draw a characteristic for a 5-level uniform quantizer with input and output range $(0, 1]$. Assume that $(0, 0.2] \to 0.2$. This quantizer has to be used in combination with a compressor and an expander to produce the quantizer of Problem 7 with $\rho = 2$. Construct a compressor function $c()$, as is done in Figure 4.10.

Problem 20 Design a first-order noise-shaper that suppresses noise at high frequencies.

Problem 21 Compute the differential entropy (4.23) for the Laplacian probability density function

$$p(x) = \frac{\alpha}{2} \exp(-\alpha|x|), \alpha > 0.$$

Express this in terms of the variance $\sigma^2 = 2/\alpha^2$. Compute $H(\hat{x}[n])$ for this distribution and compare it with the $H(\hat{x}[n])$ of (4.29).

Problem 22 Verify (4.24).

5

Signals and transformations

5.1 Introduction

The source-coding systems discussed so far consist only of quantizer, reconstructor, coder and decoder blocks. Diagrams of such source encoders and decoders are shown in Figure 5.1. It has been shown in Chapters 3 and 4 that, if quantizer-output symbols are dependent, a complex coder is required to achieve the minimum bit rate. In this chapter it is shown that under certain conditions, by using transformations, the complexity of the coder in the source encoder can be kept low. The general diagram of such a source coder is shown in Figure 5.2.

Another reason for using transformations is a perceptual one. If the proper transformation is used, quantization of the transformed signal leads to a distortion in the reconstructed signal $\hat{x}[n]$ which, for the same bit rate, is perceptually less disturbing than if the transformation is not used.

The three best-known types of transformations are introduced in separate sections. Section 5.4 discusses the use of orthogonal transforms. An example of such a transform is the well-known discrete cosine transform (DCT). Section 5.5 discusses subband filtering and Section 5.6 predictive filtering. Orthogonal transforms and

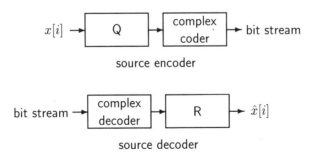

Figure 5.1 *Source-coding system without transformation.*

88 Signals and transformations

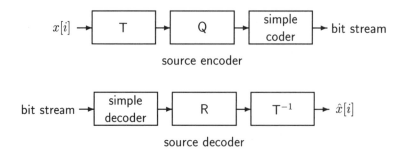

Figure 5.2 *Source-coding system with transformation.*

subband filtering are treated separately in this book. In a more recently published conceptual framework, e.g. [49], it is recognized that they can be seen as belonging to the same class of transformations. The name of the signal transformation used is usually given to the complete coding system: we speak of transform coding, subband coding and predictive coding. Specific examples of source-coding systems based on these transformations are given in Part II.

An important issue in this chapter is the entropy of a quantized signal. The discussion on the entropy of quantized signals in Chapter 4 is therefore first extended to statistically dependent sources in Section 5.2. In this section relations are established between statistical signal properties such as power spectral density and entropy for stationary Gaussian signals [10].

A warning is called for here. In this chapter, transforms are used in simple, straightforward source-coding systems. For instance, almost all quantizers used in the derivations and examples are uniform with step size Δ, and signals are assumed to be stationary and Gaussian. And, in spite of the remarks that have previously been made on its uselessness, the mean-square error is used as a distortion measure. Also, the symbol $\simeq$ is often used as if it were $=$. This leads to derivations of the form $a \simeq b$, $b < c$, therefore $a < c$. However, a more precise mathematical reasoning, which is not presented, can justify the results. The main purpose of this chapter is to illustrate why transformations are useful. This itself proves to be a difficult task, even for the simple cases discussed here.

5.2 Quantizing and coding of dependent sources

It was found in Section 4.5 that if a statistically independent source is uniformly quantized with a small step size Δ and coded with a Huffman code, a bit rate R

can be achieved that satisfies

$$H(\hat{x}[n]) \leq R < H(\hat{x}[n]) + 1. \tag{5.1}$$

The mean-square error is used as a distortion measure. The distortion D_{MSE} is given by

$$D_{\text{MSE}} = \frac{\Delta^2}{12}; \tag{5.2}$$

see (4.11). The complexity of the Huffman code is determined by the number of Huffman table entries (4.27)

$$M = K\frac{\sigma}{\Delta}. \tag{5.3}$$

The above results remain valid if the samples $x[n]$ come from a dependent source, as long as the quantizer-output symbols are coded independently. In this section expressions for the entropy, the bit rate and the coder complexity are derived for statistically dependent sources and dependently coded quantizer-output symbols.

An expression, similar to (4.22), for the entropy of quantizer-output symbols, given a step size Δ, can also be derived for the entropy of a sequence of N quantizer-output symbols from a quantized dependent source. It follows that

$$\frac{1}{N}H(\hat{x}[n],\ldots,\hat{x}[n+N-1]) \tag{5.4}$$

$$\simeq -\frac{1}{N}\int_{\mathbf{u}\in\mathbb{R}^N} p_{\mathbf{x}}(\mathbf{u})\log_2(p_{\mathbf{x}}(\mathbf{u}))d\mathbf{u} - \log_2(\Delta),$$

where $p_{\mathbf{x}}(\mathbf{u})$ is the joint probability density function of the vector $\mathbf{x} = (x[n],\ldots,x[n+N-1])^{\text{T}}$ at $\mathbf{u} = (u_1,\ldots,u_N)^{\text{T}}$. The bit rate R to code N consecutive samples satisfies

$$\frac{1}{N}H(\hat{x}[n],\ldots,\hat{x}[n+N-1]) \leq R < \frac{1}{N}H(\hat{x}[n],\ldots,\hat{x}[n+N-1]) + \frac{1}{N}. \tag{5.5}$$

The distortion is still given by (5.2). If groups of N quantizer-output symbols are coded together, the number of entries in the Huffman table follows from (4.28), which is repeated here

$$M^N = K^N \left(\frac{\sigma}{\Delta}\right)^N. \tag{5.6}$$

The number of entries in the Huffman table (5.6) increases rapidly with N. A source coder aiming to reach a bit rate satisfying (5.5) by using Huffman coding of quantizer-output symbols would require a large memory to store the Huffman table. This would make the source coder very complex. A solution to this problem will be presented later on in this chapter. First some other properties of signal entropies will be discussed.

The term
$$h(\mathbf{x}) \stackrel{\text{def}}{=} -\int_{\mathbf{u}\in\mathrm{I\!R}^N} p_\mathbf{x}(\mathbf{u})\log_2(p_\mathbf{x}(\mathbf{u}))d\mathbf{u} \tag{5.7}$$

in (5.4) is differential entropy of the vector source that produces $\mathbf{x}$. If the mean-square error is the distortion measure, (5.4) gives the entropy for a distortion (5.2).

On combination with the results of Section 3.6 it follows that

$$H_\infty(\hat{x}[n]) = \lim_{N\to\infty} \frac{1}{N} H(\hat{x}[n],\ldots,\hat{x}[n+N-1]) \simeq \frac{1}{N}h(\mathbf{x}) - \log_2(\Delta).$$

Therefore, if N is large enough, the bit rate R, expressed in bits per symbol, is given by

$$\frac{1}{N}h(\mathbf{x}) - \log_2(\Delta) \leq R < \frac{1}{N}h(\mathbf{x}) - \log_2(\Delta) + \frac{1}{N}. \tag{5.8}$$

In the remainder of this section the above results can be made more precise for Gaussian signals which are uniformly quantized with a sufficiently small step size Δ. Moreover, $H_\infty(\hat{x}[n])$ will be related to the signal's autocorrelation function and power spectral density.

The derivation for the entropy of a sequence of N quantizer-output symbols from a quantized Gaussian signal is somewhat more tedious than for the case of an independent Gaussian signal, given in (4.29). Assume that the elements of the vector $\mathbf{x} = (x[n],\ldots,x[n+N-1])^\mathrm{T}$ are N dependent Gaussian stochastic variables. The joint probability density function is then given by

$$p_\mathbf{x}(\mathbf{u}) = \frac{1}{(2\pi)^{\frac{N}{2}}\sqrt{|\mathbf{C}|}} \exp\left(-\frac{\mathbf{u}^\mathrm{T}\mathbf{C}^{-1}\mathbf{u}}{2}\right), \tag{5.9}$$

with the autocorrelation matrix $\mathbf{C}$ given by

$$\mathbf{C} = \mathcal{E}\{\mathbf{x}\mathbf{x}^\mathrm{T}\},$$

or

$$C_{i,j} = \mathcal{E}\{x_i x_j\}, \ i,j = 1,\ldots,N,$$

and $|\mathbf{C}|$ denoting the determinant of $\mathbf{C}$.

If the vector $\mathbf{x}$ contains consecutive samples from a stationary signal $x[n]$, the signal's autocorrelation function is defined by

$$R[k] = \mathcal{E}\{x[n]x[n+k]\}, \ k,n \in \mathbb{Z},$$

and

$$C_{i,j} = R[i-j], \ i,j = 1,\ldots,N.$$

The autocorrelation matrix $\mathbf{C}$ is constant on all its diagonals. A matrix of this form is called a Toeplitz matrix. In this case $\mathbf{C}$ is called the $N \times N$ autocorrelation matrix.

The correlation matrix $\mathbf{C}$ is symmetrical and positive definite. This means that there are N positive eigenvalues $\lambda_1, \ldots, \lambda_N$ and N eigenvectors $\mathbf{v}_1, \ldots, \mathbf{v}_N$, such that [11]

$$\mathbf{C}\mathbf{v}_i = \lambda_i \mathbf{v}_i, \ i = 1, \ldots, N, \tag{5.10}$$

and that

$$|\mathbf{C}| = \prod_{i=1}^{N} \lambda_i. \tag{5.11}$$

Observe also that $\sigma_x^2 = R[0] = C_{i,i}, \ i = 1, \ldots, N$. Moreover, it is well known [11] that

$$\sum_{i=1}^{N} C_{i,i} = \sum_{i=1}^{N} \lambda_i = N\sigma_x^2.$$

Therefore,

$$\sigma_x^2 = \frac{1}{N} \sum_{i=1}^{N} \lambda_i. \tag{5.12}$$

The entropy (4.29), expressed in the eigenvalues of the autocorrelation matrix, is then given by

$$H(\hat{x}[n]) \simeq \frac{1}{2} \log_2(2\pi e) + \log_2(\frac{\sqrt{\frac{1}{N}\sum_{i=1}^{N} \lambda_i}}{\Delta}). \tag{5.13}$$

It can be shown that for a Gaussian vector $\hat{\mathbf{x}}$ of which the components are uniformly quantized with a step size Δ

$$H(\hat{x}_1, \ldots, \hat{x}_N) \tag{5.14}$$

$$\simeq \frac{N}{2} \log_2(2\pi e) + \log_2(\frac{\sqrt{|\mathbf{C}|}}{\Delta^N})$$

$$= \frac{N}{2} \log_2(2\pi e) + \log_2(\frac{\sqrt{\prod_{i=1}^{N} \lambda_i}}{\Delta^N})$$

$$= \frac{N}{2} \log_2(2\pi e) + \sum_{i=1}^{N} \log_2(\frac{\sqrt{\lambda_i}}{\Delta}).$$

To avoid a complicated mathematical analysis, it is assumed that

$$\Delta \ll \min_{i=1,\ldots,N}(\sqrt{\lambda_i}).$$

If $\mathbf{x}$ denotes a vector of N consecutive samples of a signal $x[n]$, the entropy per symbol of a sequence of N consecutive quantizer-output symbols is given by

$$\frac{1}{N} H(\hat{x}[n], \ldots, \hat{x}[n+N-1]) \tag{5.15}$$

$$\simeq \frac{1}{2} \log_2(2\pi e) + \frac{1}{N} \sum_{i=1}^{N} \log_2(\frac{\sqrt{\lambda_i}}{\Delta}).$$

This expression relates the entropy of the quantizer-output symbols of a quantized Gaussian signal to the eigenvalues of the autocorrelation matrix of $x[n]$ and the step size Δ.

The minimum bit rate for a Gaussian stationary signal quantized with a step size Δ is given by

$$H_\infty(\hat{x}[n]) = \lim_{N \to \infty} \frac{1}{N} H(\hat{x}[n], \ldots, \hat{x}[n+N-1]).$$

In the case of stationary Gaussian signals, $H_\infty(\hat{x}[n])$ can be related to the signal's power spectral density $S(\theta)$, defined by [10, 11, 13]

$$S(\theta) = \sum_{k=-\infty}^{+\infty} R[k] \exp(jk\theta).$$

For positive definite Toeplitz matrices the Szegö limit theorem [50] applies. According to this theorem,

$$\lim_{N \to \infty} \frac{1}{N} \sum_{i=1}^{N} \log_2(\sqrt{\lambda_i}) = \frac{1}{2\pi} \int_{-\pi}^{+\pi} \log_2(\sqrt{S(\theta)}) d\theta.$$

In combination with (3.22) and (5.15) this gives

$$H_\infty(\hat{x}[n]) \tag{5.16}$$
$$= \lim_{N \to \infty} \frac{1}{N} H(\hat{x}[n], \ldots, \hat{x}[n+N-1])$$
$$\simeq \frac{1}{2} \log_2(2\pi e) + \lim_{N \to \infty} \frac{1}{N} \sum_{i=1}^{N} \log_2\left(\frac{\sqrt{\lambda_i}}{\Delta}\right)$$
$$= \frac{1}{2} \log_2(2\pi e) + \frac{1}{2\pi} \int_{-\pi}^{+\pi} \frac{1}{2} \log_2(S(\theta)) d\theta - \log_2(\Delta).$$

This expression brings out the relation between the entropy of the quantizer-output symbols of a quantized Gaussian signal and the signal's power spectral density and the step size Δ.

In practice, power spectral densities are often assumed rational, which means that

$$S(\theta) = \sigma_\epsilon^2 \frac{|B(\exp(j\theta))|^2}{|A(\exp(j\theta))|^2}, \tag{5.17}$$

with $A(z)$ and $B(z)$ of the form

$$A(z) = 1 + a_1 z^{-1} + \ldots + a_p z^{-p},$$

and

$$B(z) = 1 + b_1 z^{-1} + \ldots + b_p z^{-p},$$

and σ_ϵ^2 a constant. Furthermore, $A(z)$ and $B(z)$ are chosen such that $z^p A(z)$ and $z^p B(z)$ only have zeros inside the unit circle of the complex plane[1]. With some complex analysis it can be shown that

$$\frac{1}{2\pi} \int_{-\pi}^{\pi} \log_2\left(\sqrt{S(\theta)}\right) d\theta \tag{5.18}$$

$$= \frac{1}{2\pi} \int_{-\pi}^{\pi} \log_2\left(\sqrt{\sigma_\epsilon^2 \frac{|B(\exp(j\theta))|^2}{|A(\exp(j\theta))|^2}}\right) d\theta \tag{5.19}$$

$$= \log_2(\sigma_\epsilon).$$

On substitution of this result in (5.16), and assuming that $\Delta \ll \sigma_\epsilon$, we obtain

$$H_\infty(\hat{x}[n]) \simeq \frac{1}{2}\log_2(2\pi e) + \log_2\left(\frac{\sigma_\epsilon}{\Delta}\right). \tag{5.20}$$

It is interesting to note that for a given correlation matrix the Gaussian probability distribution function (5.9) maximizes the differential entropy (5.7).

5.3 Relations to the rate-distortion theory

As was the case with the results of Section 4.5, the results of Section 5.2 can also be related to the rate-distortion theory [6, 7]. As in Section 4.6, it is assumed here that the signals are Gaussian and that the distortion measure is the mean-square error D_{MSE} of (1.4). No derivations but only results are presented.

The bit rate, expressed in bits per input symbol, of a source-coding system at a given distortion D is denoted by $R(D)$. For each input probability density function $p_\mathbf{x}(\mathbf{x})$ there exists a bound, the rate-distortion bound $R_B(D)$, such that (4.30) holds. Here, $R_B(D)$ is given for dependent, stationary input samples with a power spectral density $S(\theta)$. The result is given in a parametric form, in which ψ is a parameter traversing the interval

$$[0, \max_\theta(S(\theta))];$$

see [6]. Every value of ψ in this interval corresponds to a point on the rate-distortion bound. The distortion for a certain ψ is given by

$$D_\psi = \frac{1}{2\pi} \int_{-\pi}^{\pi} \min(\psi, S(\theta)) \, d\theta, \tag{5.21}$$

and the rate by

$$R_B(D_\psi) = \frac{1}{2\pi} \int_{-\pi}^{+\pi} \max\left(0, \frac{1}{2}\log_2\left(\frac{S(\theta)}{\psi}\right)\right) d\theta. \tag{5.22}$$

---

[1] For notational simplicity the orders of $A(z)$ and $B(z)$ are chosen to be identical.

94 Signals and transformations

The rate-distortion bound can be constructed by computing all points $(D_\psi, R(D_\psi))$ for values of ψ in the interval

$$[0, \max_\theta(S(\theta))].$$

Figure 5.3 shows an example of a signal's power spectral density. The horizontal line has a height ψ. Expressions (5.21) and (5.22) state that the parts of the signal's power spectral density below this line determine the distortion and the parts above it the bit rate.

If for all θ

$$D < \min_\theta(S(\theta)), \tag{5.23}$$

then

$$R_B(D) = \frac{1}{2\pi} \int_{-\pi}^{+\pi} \frac{1}{2} \log_2(\frac{S(\theta)}{D}) d\theta. \tag{5.24}$$

In this case, in a similar manner as for statistically independent sources, we obtain for dependent stationary sources

$$H_\infty(\hat{x}[n]) \simeq R_B(D) + \frac{1}{2} \log_2\left(\frac{e\pi}{6}\right) \simeq R_B(D) + 0.25. \tag{5.25}$$

This result is identical to (4.32) for independent sources. The condition (5.23) must be added to the remarks that follow the derivation of (4.32). What has been obtained here is a Gish-Pierce asymptote [30, 41] for dependent Gaussian sources.

In the following example the rate-distortion bound is computed for the second-order autoregressive process of Figure 5.3. This example will also be used in the next section to illustrate the usefulness of orthogonal transforms in source-coding systems.

Example 19 An autoregressive process $x[n]$ is generated passing a white-noise sequence $e[n]$ through an all-pole filter. The $x[n]$ satisfy

$$x[n] + a_1 x[n-1] + \ldots + a_p x[n-p] = e[n]. \tag{5.26}$$

The integer p is the order of the process and the coefficients $a_1, \ldots, a_p$ are the autoregressive parameters. The white noise signal $e[n]$ is called the excitation signal, it has variance σ_e^2 and zero mean. In the z-domain $x[n]$ is given by

$$X(z) = \frac{E(z)}{1 + a_1 z^{-1} + \ldots + a_p z^{-p}} = \frac{E(z)}{A(z)}. \tag{5.27}$$

The power spectral density function is given by

$$S(\theta) = \frac{\sigma_e^2}{|A(\exp(j\theta))|^2}, \tag{5.28}$$

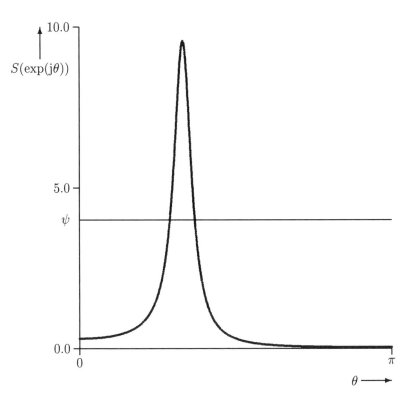

Figure 5.3 *Power spectral density function of a second-order autoregressive process. The horizontal line, with height ψ, determines the integrands of (5.21) and (5.22).*

from which it is clear that an autoregressive process is an example of a signal with a rational power spectral density the general form of which is given in (5.17). It is further assumed that the $e[n]$ are Gaussian.

In this example $p = 2$ and

$$a_1 = -2\rho\cos(\Omega) \qquad (5.29)$$
$$a_2 = \rho^2. \qquad (5.30)$$

This process has poles at locations $(\rho\cos(\Omega), \pm j\rho\sin(\Omega))$. Figure 5.3 shows the power spectral density for $\rho = 0.9$ and $\Omega = \pi/3$. The excitation signal's variance σ_e^2 is chosen such that $x[n]$ has unit variance. This implies that

$$\sigma_e^2 = 0.259.$$

The rate-distortion bound for these parameters is computed by using (5.21) and (5.22) and is shown in Figure 5.4. According to (5.23) and (5.24), the asymptote for small D is

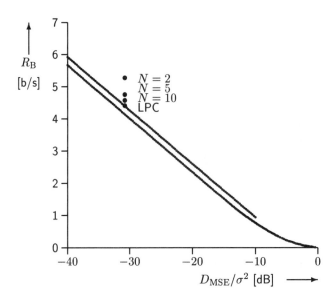

Figure 5.4 Rate-distortion function (lower curve) and Gish-Pierce bound (upper curve) for an autoregressive process of order 2, with $a_1 = -0.9$, $a_2 = 0.81$. Lower bounds to bit rate for transform coder (KLT) with several block lengths N and for a linear predictive coder LPC.

given by

$$\frac{1}{2}\log_2(\frac{\sigma_e^2}{D}).$$

In this example the asymptote intersects the distortion axis at $D = 0.259$, which corresponds to -5.86 dB.

If $x[n]$ is quantized with a step size $\Delta \ll \sigma_e$, then

$$H_\infty(\hat{x}[n]) \simeq \frac{1}{2}\log_2(2\pi e) + \log_2\left(\frac{\sigma_e}{\Delta}\right) \simeq \frac{1}{2}\log_2(\frac{\sigma_e^2}{D}) + 0.25. \quad (5.31)$$

This is the Gish-Pierce asymptote for this example, which is also shown in Figure 5.4.
◊

5.4 Orthogonal transforms

5.4.1 Introduction

Orthogonal transforms are block-based operations. The input signal $x[n]$ is divided into blocks of length N, which are further treated as vectors $\mathbf{x}$. Sometimes the input

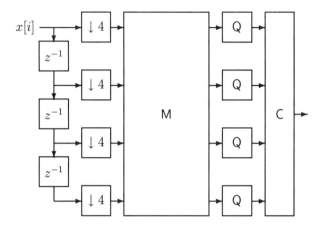

Figure 5.5 *Transform encoder with a block length of four.*

signal is already presented in a vector form. For example, in the case of stereophonic audio signals a sample of the left and the right channel can be combined to a vector of length two [51]. The vector **x** is transformed to another representation, denoted by **y**, by multiplication with a transform matrix **M** such that

$$\mathbf{y} = \mathbf{M}\mathbf{x}. \tag{5.32}$$

The matrix **M** is orthogonal, therefore its inverse is given by

$$\mathbf{M}^{-1} = \mathbf{M}^\mathrm{T}. \tag{5.33}$$

The superscript T denotes matrix or vector transposition. The elements of the vector **y** are quantized and coded. The elements of the vector **y** are called transform coefficients. In the source decoder the quantized vector elements are reconstructed as $\hat{\mathbf{y}}$ and a reconstruction of **x** is computed as

$$\hat{\mathbf{x}} = \mathbf{M}^\mathrm{T}\hat{\mathbf{y}}. \tag{5.34}$$

Examples of a transform encoder and decoder with block lengths of four are given in Figures 5.5 and 5.6, respectively. Boxes M and M^T perform the orthogonal transform and its inverse. From these pictures it is clear that the APCM coding of Figures 2.2 on page 22 and 2.3 on page 23 can be seen as a special case of transform coding with **M** replaced by a diagonal matrix.

If $\mathbf{e}_y$ is the quantization error in **y**, then

$$\mathbf{e}_y = \hat{\mathbf{y}} - \mathbf{y}. \tag{5.35}$$

This results in an error $\mathbf{e}_x$ in the output, given by

$$\mathbf{e}_x = \hat{\mathbf{x}} - \mathbf{x} = \mathbf{M}^\mathrm{T}\mathbf{e}_y. \tag{5.36}$$

98 Signals and transformations

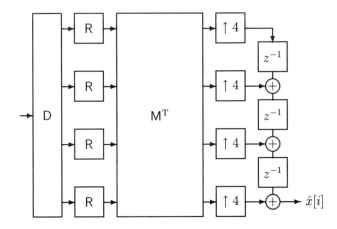

Figure 5.6 *Transform decoder with a block length of four.*

Figure 5.7 is the transform coding version of Figure 4.2 on page 53. If D_{MSE} from (1.4) is the distortion measure, then

$$\begin{aligned}
D_{\text{MSE}} &= \frac{1}{N}\mathcal{E}\{(\hat{\mathbf{x}} - \mathbf{x})^T(\hat{\mathbf{x}} - \mathbf{x})\} \\
&= \frac{1}{N}\mathcal{E}\{\mathbf{e}_y^T \mathbf{M}^T \mathbf{M} \mathbf{e}_y\} \\
&= \frac{1}{N}\mathcal{E}\{\mathbf{e}_y^T \mathbf{e}_y\}.
\end{aligned} \qquad (5.37)$$

This shows that the mean-square error is unaffected by an orthogonal transform. The correlation matrix of the coding error is given by

$$\begin{aligned}
\mathbf{C}_e &= \mathcal{E}\{(\hat{\mathbf{x}} - \mathbf{x})(\hat{\mathbf{x}} - \mathbf{x})^T\} \\
&= \mathcal{E}\{\mathbf{M}^T \mathbf{e}_y \mathbf{e}_y^T \mathbf{M}\} \\
&= \mathbf{M}^T \mathcal{E}\{\mathbf{e}_y \mathbf{e}_y^T\} \mathbf{M}.
\end{aligned} \qquad (5.38)$$

If $\mathbf{y}$ is uniformly quantized with step size Δ, the quantization error is uncorrelated with variance

$$D_{\text{MSE}} = \frac{\Delta^2}{12}. \qquad (5.39)$$

Therefore

$$\mathbf{C}_e = \frac{\Delta^2}{12}\mathbf{I}. \qquad (5.40)$$

If the input signal is two-dimensional, such as a picture, it is divided into blocks of $N_v \times N_h$ samples. The N_v and N_h are the block's dimensions in the vertical

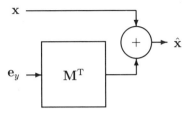

Figure 5.7 *Additive source-decoding model for transform coders.*

and horizontal directions, respectively. The blocks are further denoted as $N_v \times N_h$ matrices $\mathbf{X}$. The matrix $\mathbf{X}$ is transformed to the matrix $\mathbf{Y}$ by pre- and post-multiplying it by transform matrices $\mathbf{M}_v$ and $\mathbf{M}_h^T$ such that

$$\mathbf{Y} = \mathbf{M}_v \, \mathbf{X} \, \mathbf{M}_h^T.$$

The matrix $\mathbf{M}_v$ is an $N_v \times N_v$ matrix transforming the columns of $\mathbf{X}$, while $\mathbf{M}_h$ is an $N_h \times N_h$ matrix transforming the rows of $\mathbf{X}$. In the source decoder a reconstruction $\hat{\mathbf{X}}$ of $\mathbf{X}$ is obtained from the reconstructed block $\hat{\mathbf{Y}}$ with

$$\hat{\mathbf{X}} = \mathbf{M}_v^T \, \hat{\mathbf{Y}} \, \mathbf{M}_h.$$

Popular dimensions in picture coding for the blocks $\mathbf{X}$ are 8×8 and 16×16. A popular transform for picture coding is the discrete cosine transform (DCT) [4, 52]. Other transforms are the Walsh-Hadamard transform (WHT), once popular because the transform matrix only contains numbers 1 and -1 and allowed implementations of acceptable complexity, and the discrete Fourier transform (DFT). Many transforms are discussed in [52]. In the remainder of this section only one-dimensional transforms are considered.

There are two reasons for using orthogonal transforms. The first is that a transform can make it easier to reduce the bit rate by coding quantizer-output symbols. That this is optimally done by the proper orthogonal transform will be illustrated for stationary Gaussian signals with the coding-error variance as the distortion measure. The second reason to apply transforms is that a transform can decompose the signal into perceptually relevant components. By quantizing these components with different accuracies, the coding errors can be controlled in such a way that perceptible distortion is minimal.

5.4.2 Transform coding and bit rate

The following example illustrates for the simple case of $N = 2$ how transforming a signal makes coding easier. First the number of bits needed to code the components

100 Signals and transformations

of the quantized input vector **x** separately is computed. Then it is shown that this number can be reduced if the vector is first transformed.

Example 20 Assume that the components of the vector **x** have a joint Gaussian probability density function [10]

$$p_{\mathbf{x}}(\mathbf{u}) = \frac{1}{2\pi\sqrt{|\mathbf{C}|}} \exp\left(-\frac{\mathbf{u}^{\mathrm{T}}\mathbf{C}^{-1}\mathbf{u}}{2}\right). \tag{5.41}$$

The 2×2 matrix **C** is the correlation matrix, which is defined by

$$\mathbf{C} = \begin{pmatrix} \mathcal{E}\{x_1^2\} & \mathcal{E}\{x_1 x_2\} \\ \mathcal{E}\{x_1 x_2\} & \mathcal{E}\{x_2^2\} \end{pmatrix}.$$

In the above definitions it has been assumed that $\mathcal{E}\{x_i\} = 0$, $i = 1, 2$. The determinant of **C** is denoted by $|\mathbf{C}|$.

In this example it is also assumed that

$$\mathcal{E}\{x_1^2\} = \mathcal{E}\{x_2^2\} = \sigma_x^2,$$

and that

$$\mathcal{E}\{x_1 x_2\} = \rho \sigma_x^2.$$

Therefore,

$$\mathbf{C} = \sigma_x^2 \begin{pmatrix} 1 & \rho \\ \rho & 1 \end{pmatrix},$$

and

$$|\mathbf{C}| = \sigma_x^4 (1 - \rho^2).$$

A contour plot of $p_{\mathbf{x}}(\mathbf{u})$ for $\sigma_x^2 = 0.25$ and $\rho = 0.9$ is shown in Figure 5.8. The contours are shown for $\log(p_{\mathbf{x}}(\mathbf{u})) = 0.25,\ 0,\ -0.25,\ -0.50$.

The probability density function $p_{x_1}(u_1)$ of x_1 follows from [10]

$$p_{x_1}(u_1) = \int_{-\infty}^{+\infty} p_{\mathbf{x}}(\mathbf{u})\, du_2.$$

From evaluating the above integral it follows that

$$p_{x_1}(u) = p_{x_2}(u) = \frac{1}{\sqrt{2\pi\sigma_x^2}} \exp\left(-\frac{u^2}{2\sigma_x^2}\right).$$

Assume first in this example that x_1 and x_2 are quantized separately by a uniform midtread quantizer with step size Δ. The quantization levels are shown in Figure 5.8 for $\Delta = 0.1$. It has been shown in Section 5.2 that the entropy of the quantizer-output symbols is closely approximated by

$$H(\hat{x}_1) = H(\hat{x}_2) \simeq \frac{1}{2}\log_2(2\pi e) + \log_2\left(\frac{\sigma_x}{\Delta}\right). \tag{5.42}$$

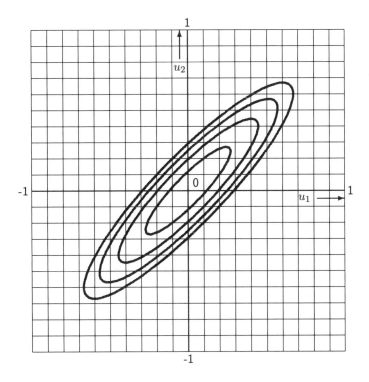

Figure 5.8 *Contour plot of a joint Gaussian probability density function.*

Independent coding of the $\hat{x}_1$ and $\hat{x}_2$ leads to an average bit rate R satisfying

$$\frac{1}{2}(H(\hat{x}_1) + H(\hat{x}_1)) \leq R < \frac{1}{2}(H(\hat{x}_1) + H(\hat{x}_1)) + 1, \tag{5.43}$$

and a quantization-error variance σ_e^2 is given by

$$\sigma_e^2 = \frac{\Delta^2}{12}. \tag{5.44}$$

Now the situation is considered where the components of a transformed vector

$$\mathbf{y} = \mathbf{M}\mathbf{x} = \frac{1}{\sqrt{2}} \begin{pmatrix} 1 & 1 \\ 1 & -1 \end{pmatrix} \mathbf{x}$$

are quantized with the same quantizer. The result will be that the same coding-error variance can be obtained at a lower bit rate. At the receiver the $\hat{\mathbf{x}}$ is obtained with

$$\hat{\mathbf{x}} = \mathbf{M}^T \hat{\mathbf{y}} = \frac{1}{\sqrt{2}} \begin{pmatrix} 1 & 1 \\ 1 & -1 \end{pmatrix} \hat{\mathbf{y}}.$$

According to (5.39), the variance of the coding error per sample is given by

$$\frac{\Delta^2}{12},$$

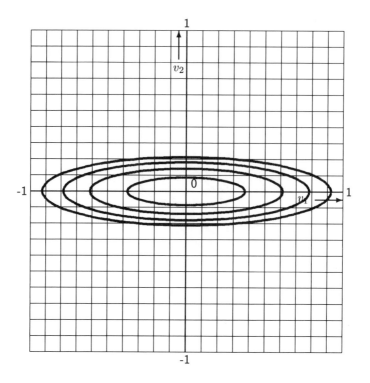

Figure 5.9 *Contour plot of a joint Gaussian probability density function after an orthogonal transform.*

which is the same as without the transform.
Because

$$\mathcal{E}\{\mathbf{y}\mathbf{y}^T\} = \mathbf{M}\,\mathbf{C}\,\mathbf{M}^T = \sigma_x^2 \begin{pmatrix} 1+\rho & 0 \\ 0 & 1-\rho \end{pmatrix},$$

the y_1 and y_2 are statistically independent. The joint probability density function $p_\mathbf{y}(\mathbf{v})$ is given by

$$p_\mathbf{y}(\mathbf{v}) = p_{y_1}(v_1) p_{y_2}(v_2), \tag{5.45}$$

with

$$p_{y_1}(v_1) = \frac{1}{\sqrt{2\pi(1+\rho)\sigma_x^2}} \exp\left(-\frac{v_1^2}{2(1+\rho)\sigma_x^2}\right) \tag{5.46}$$

$$p_{y_2}(v_2) = \frac{1}{\sqrt{2\pi(1-\rho)\sigma_x^2}} \exp\left(-\frac{v_2^2}{2(1-\rho)\sigma_x^2}\right). \tag{5.47}$$

A contour plot of $p_\mathbf{y}(\mathbf{v})$ for $\sigma_x^2 = 0.25$ and $\rho = 0.9$ for $p_\mathbf{y}(\mathbf{v})$ is shown in Figure 5.9. The contours are shown for $\log(p_\mathbf{y}(\mathbf{v})) = 0.25,\ 0,\ -0.25,\ -0.50$.

Instead of the components of **x** the components of **y** are quantized, with the same step size Δ, and coded. The quantization levels are shown in Figure 5.9 for the case of $\Delta = 0.1$. The entropies of the quantizer-output symbols are given by

$$H(\hat{y}_1) \simeq \frac{1}{2}\log_2(2\pi e) + \log_2(\frac{\sqrt{(1+\rho)}\sigma_x}{\Delta})$$

and

$$H(\hat{y}_2) \simeq \frac{1}{2}\log_2(2\pi e) + \log_2(\frac{(\sqrt{1-\rho})\sigma_x}{\Delta}).$$

The entropy per sample, determining the number of bits per sample needed to code the signal, is given by

$$\frac{1}{2}H(\hat{y}_1, \hat{y}_2) = \frac{1}{2}(H(\hat{y}_1) + H(\hat{y}_2))$$
$$\simeq \frac{1}{2}\log_2(2\pi e) + \log_2(\frac{\sigma_x}{\Delta}) + \log_2(\sqrt{1-\rho^2}). \tag{5.48}$$

The term $\log_2(\sqrt{1-\rho^2})$ is negative and reduces the entropy per sample with respect to (5.42). In this example $\rho = 0.9$, which leads to a reduction of 1.2 bits per sample for an unchanged coding error.

Independent coding of the $\hat{y}_1$ and $\hat{y}_2$ leads to an average bit rate R which satisfies

$$\frac{1}{2}(H(\hat{y}_1) + H(\hat{y}_1)) \le R < \frac{1}{2}(H(\hat{y}_1) + H(\hat{y}_1)) + 1. \tag{5.49}$$

Combined coding of the $\hat{y}_1$ and $\hat{y}_2$ reduces the upper bound of (5.49) by $1/2$.

To assess the practical value of transforms, four situations are compared. The first is independent coding of the quantized x_1 and x_2, the second is joint coding of the quantized x_1 and x_2, the third is independent coding of the quantized y_1 and y_2 and the fourth is joint coding of the quantized y_1 and y_2. It is assumed that $\Delta = \sigma_x/10$ and $\rho = 0.9$.

First consider independent coding of the quantized x_1 and x_2. Assume that the input range of the quantizer equals $(-4\sigma_x, 4\sigma_x)$. The number of quantization levels then equals 81. Overload distortion in this case is negligible. The distortion is then given by

$$\frac{\Delta^2}{12} = \frac{\sigma_x^2}{1200}$$

From (5.42) it follows that a Huffman code can be found such that the bit rate R satisfies

$$5.37 \le R < 6.37.$$

A Huffman code with 81 entries is required.

According to (5.14), joint coding of the quantized x_1 and x_2 leads to a bit rate R that satisfies

$$4.17 \le R < 4.67.$$

The distortion is the same as in the case of independent coding. However, with the given input range, a Huffman code with $81^2 = 6561$ entries is now required.

		independent coding	joint coding
QC	bit rate	$5.37 \leq R < 6.36$	$4.17 \leq R < 4.67$
	entries	81	6561
TQC	bit rate	$4.17 \leq R < 5.17$	$4.17 \leq R < 4.67$
	entries	138	2912

Table 5.1 *Bit rates and numbers of entries of Huffman tables for coders with and without an orthogonal transform.*

Independent coding of quantized y_1 and y_2 leads to a bit rate R satisfying

$$4.17 \leq R < 5.17,$$

again for the same distortion. Assume that the input ranges of the y_1 and the y_2 quantizers are equal to

$$(-4\sqrt{1+\rho}\,\sigma_x, 4\sqrt{1+\rho}\,\sigma_x)$$

and

$$(-4\sqrt{1-\rho}\,\sigma_x, 4\sqrt{1-\rho}\,\sigma_x),$$

respectively, in order to have the same negligible overload probability as in the previous cases. In this case two Huffman codes are required, one for the quantized y_1 and one for the quantized y_2. The numbers of entries are 112 and 26, respectively. The total number of entries is 138.

Joint coding of quantized y_1 and y_2 can reduce the upper bound to the bit rate to 4.67, in which case one Huffman code with $112 \times 26 = 2912$ entries is required.

From the above comparison it is clear that an orthogonal transform before quantization greatly reduces coder complexity. The results are summarized in Table 5.1.
◇

In Example 20 two methods of coding a vector **x** have been compared. The first consists of the independent quantization and coding of the two components of the vector **x**. In the second method an orthogonal transform is first applied to the vector. Subsequently the components of the transformed vector are quantized independently and coded. It has been observed that the components of the transformed vector **y** were independent. The conclusion drawn in Example 20 is that the latter method leads to a result that can be coded with a smaller number of bits for the same coding error. Note that in Example 20

$$\mathbf{M\,C\,M}^T = \sigma_x^2 \begin{pmatrix} 1+\rho & 0 \\ 0 & 1-\rho \end{pmatrix}.$$

The transform matrix **M** diagonalizes the correlation matrix **C**.

The above results can be generalized to larger vectors. The case considered is that of a stationary Gaussian source producing samples $x[n]$, which have zero mean, variance σ_x^2 and autocorrelation function $R_{xx}[k]$. A property used throughout is that, if a vector with a joint Gaussian probability density function is transformed, the resulting vector also has a joint Gaussian probability density function.

Before quantization the vectors are multiplied by a matrix $\mathbf{M}$ that diagonalizes $\mathbf{C}$ such that

$$\mathbf{MCM}^T = \begin{pmatrix} \lambda_1 & & \\ & \ddots & \\ & & \lambda_N \end{pmatrix}.$$

The $\lambda_1, \ldots, \lambda_N$ are the eigenvalues of $\mathbf{C}$ and the columns of $\mathbf{M}^T$ are the eigenvectors of $\mathbf{C}$. Let

$$\mathbf{y} = \mathbf{Mx}.$$

The correlation matrix of $\mathbf{y}$ is given by

$$\mathcal{E}\{\mathbf{yy}^T\} = \mathbf{M}\mathcal{E}\{\mathbf{xx}^T\}\mathbf{M}^T = \mathbf{MCM}^T = \begin{pmatrix} \lambda_1 & & \\ & \ddots & \\ & & \lambda_N \end{pmatrix},$$

which shows that components of $\mathbf{y}$ are uncorrelated. Because they are Gaussian, they are also independent [10]. The variance of y_i is given by λ_i.

If the components of the vector $\mathbf{y}$ are quantized with a uniform quantizer with step size Δ and coded, then each quantized component can be considered as the output of a discrete source. After decoding the distorted vector $\hat{\mathbf{x}}$ is obtained from

$$\hat{\mathbf{x}} = \mathbf{M}^T \hat{\mathbf{y}}.$$

It has been shown in Example 20 that D still equals $\frac{\Delta^2}{12}$. The entropy of the quantizer-output symbols obtained by quantizing y_i is given by

$$H(\hat{y}_i) \simeq \frac{1}{2}\log_2(2\pi e) + \log_2(\frac{\sqrt{\lambda_i}}{\Delta}). \tag{5.50}$$

The average entropy per transform coefficient follows from

$$\frac{1}{N}\sum_{i=1}^{N} H(\hat{y}_i) \simeq \frac{1}{2}\log_2(2\pi e) + \frac{1}{N}\sum_{i=1}^{N} \log_2(\frac{\sqrt{\lambda_i}}{\Delta}), \tag{5.51}$$

which is identical to the average entropy per sample of the untransformed quantized signal given by (5.15), but here N independent discrete sources have to be coded, which may require different codes. Independent coding of the quantized y_n leads to a bit rate R satisfying

$$\frac{1}{N}\sum_{i=1}^{N} H(\hat{y}_i) \leq R < \frac{1}{N}\sum_{i=1}^{N} H(\hat{y}_i) + 1. \tag{5.52}$$

The upper bound to the rate can be reduced by combined coding of the quantized y_i, which leads to a bit rate R satisfying

$$\frac{1}{N}\sum_{i=1}^{N} H(\hat{y}_i) \leq R < \frac{1}{N}\sum_{i=1}^{N} H(\hat{y}_i) + \frac{1}{N}, \qquad (5.53)$$

which bounds are identical to (5.5).

It is interesting to compare coder complexity with and without the use of an orthogonal transform. Assume that, as in (4.27), overload is negligible if the number of quantization levels is given by

$$M = K\frac{\sigma_x}{\Delta} \qquad (5.54)$$

for the x_n and

$$M_i = K\frac{\sqrt{\lambda_i}}{\Delta} \qquad (5.55)$$

for the y_i. For Gaussian x_n, overload is negligible for $K \geq 8$. Combined coding of the quantized x_n leads to a bit rate (5.5) but requires a Huffman code with

$$K^N \left(\frac{\sigma_x}{\Delta}\right)^N \qquad (5.56)$$

entries. Independent coding of the quantized y_n leads to a bit rate (5.52) but requires N Huffman codes with only

$$\sum_{i=1}^{N} M_i = K\sum_{i=1}^{N} \frac{\sqrt{\lambda_i}}{\Delta} \qquad (5.57)$$

entries in all, which is substantially less than the number given by (5.56). It can be shown that the total average number of entries satisfies

$$\frac{1}{N}\sum_{i=1}^{N} M_i \leq K\frac{\sigma_x}{\Delta} \qquad (5.58)$$

and that (5.58) only holds with equality if the x_i have identical variances. Note that the average complexity is smaller than or equal to the complexity of independent Huffman coding of untransformed samples, whereas the bit rate now approximates the bit rate that can be achieved by combined Huffman coding of N untransformed samples.

The upper bound to the rate can be reduced to that in (5.5) by combined coding of the quantized y_i. This requires one Huffman code with

$$K^N \prod_{i=1}^{N} \frac{\sqrt{\lambda_i}}{\Delta} \qquad (5.59)$$

entries. From (5.12) and the fact that for $\lambda_i \geq 0$

$$\ln\left(\frac{1}{N}\sum_{i=1}^{N}\lambda_i\right) \geq \frac{1}{N}\sum_{i=1}^{N}\ln(\lambda_i), \qquad (5.60)$$

it follows that

$$K^N \prod_{i=1}^{N} \frac{\sqrt{\lambda_i}}{\Delta} < K^N \left(\frac{\sigma_x}{\Delta}\right)^N, \qquad (5.61)$$

and that (5.61) only holds with equality if the x_i are independent. From this it follows that even combined coding of the quantized y_i results in a reduction of coder complexity.

The following important conclusions can be drawn. For a stationary correlated Gaussian source there exists an orthogonal transform that converts a sequence of N dependent samples into a sequence of N independent samples, the entropy of which, after they have been uniformly quantized with step size Δ, equals the entropy per symbol of groups of N consecutive samples of the source signal quantized with the same step size. The coder complexity required to code the quantized transformed samples at a minimum bit rate is lower, usually substantially lower, than the complexity required to code the quantized untransformed samples at the same bit rate. Furthermore, the average entropy of the transformed and quantized signal can be brought arbitrarily close to $H_\infty(\hat{x}[n])$ by choosing N large enough.

The transform in the above example and derivation diagonalizes the signal's autocorrelation matrix. The name of this transform is the Karhunen-Loève transform (KLT). It is signal-dependent. It is optimal in the sense that for a given step size Δ and block length N there is no transform that further reduces the sum of the entropies of the transform coefficients, simply because it already satisfies (3.19) with equality.

The following example is an extension of Example 19. It shows how an appropriate transform helps to code the second-order autoregressive process of that example.

Example 21 The samples $x[n]$ of the autoregressive process defined in Example 19 satisfy

$$x[n] = -a_1 x[n-1] - a_2 x[n-2] + e[n],$$

with $e[n]$ independent of $x[n-1]$ and $x[n-2]$. If the $x[n]$ are uniformly quantized with step size Δ, it can be shown that the sequence of quantizer-output symbols $q[i]$ is a second-order Markov process for which probabilities $p_{j|kl}$, defined in Section 3.6, can be determined. As has been remarked in Section 3.6, an efficient method of coding such a Markov process is to use conditional Huffman codes. This means that a Huffman code is required for every combination of quantizer-output symbols $q[i-1], q[i-2]$. Assume that $\Delta = 0.1$ and $\sigma = 1$ and that a 127-level uniform midtread quantizer is used. This corresponds to a K of about 13 in (5.54). In that case this type of encoding would require $127 \times 127 \simeq 16000$ Huffman tables with 127 entries. Fortunately, because of symmetries in the input distribution and because some of the Huffman tables will be identical or almost identical, a smaller number of tables will also be sufficient, but it will still be large. Coding groups of N quantizer-output symbols together also leads to a high coder complexity, as it requires one Huffman table with a number of entries of the order of 127^N. It will now be shown that application of a Karhunen-Loève transform and uniform quantization of the transform coefficients with step size Δ, followed by independent coding of the N quantizer-output symbols, can greatly reduce the bit rate.

From (5.51), (5.52) and (5.11) it follows that the bit rate in that case satisfies

$$\frac{1}{2}\log_2(2\pi e) + \frac{1}{N}\log_2(\frac{\sqrt{|C|}}{\Delta^N}) \leq R < \frac{1}{2}\log_2(2\pi e) + \frac{1}{N}\log_2(\frac{\sqrt{|C|}}{\Delta^N}) + 1. \qquad (5.62)$$

According to (5.58), this requires N Huffman codes, with on average about 130 entries.

The bit rate can be evaluated more precisely. For this example of a second-order autoregressive process it can be shown that for $N \geq 2$

$$|C| = \frac{1}{(1-\rho^2)^2} \frac{1}{1 - 2\rho^2 \cos(2\Omega) + \rho^4} \sigma_e^{2N}.$$

This result is used in (5.62) to compute the lower bound to R. As in Example 19 we have that $\Delta = 0.1$, $\rho = 0.9$, $\Omega = \pi/3$, $\sigma = 1$ and $\sigma_e = 0.508$. This leads to a distortion

$$D = \frac{\Delta^2}{12} \frac{1}{\sigma^2}$$

of -30.8 dB and to a lower bound to the bit rate R_L, given by

$$R_L = \frac{1}{2}\log_2(2\pi e) + \log_2(\frac{\sigma_e}{\Delta}) + \frac{1.74}{N} = 4.39 + \frac{1.74}{N}$$

First note that for $N \to \infty$ the Gish-Pierce bound is approximated. Pairs (R_L, D) are shown in Figure 5.4 for several values of N. From this figure it follows that the Gish-Pierce asymptote is rapidly approached if N increases.

◊

5.4.3 Suboptimal transforms and power concentration

If the signal statistics are known to the source encoder and decoder, the Karhunen-Loève transform can be used. In general this is not the case. A possible solution would be to transmit the transform matrix regularly, but that would increase the bit rate excessively. Therefore, suboptimal transforms with a fixed transform matrix are used practically always. The most popular transform used nowadays is the discrete cosine transform (DCT). It can be shown that for large N the discrete cosine transform approximates the behaviour of the Karhunen-Loève transform. This is sometimes used as an explanation for the success of the discrete cosine transform in picture coding, even though the assumptions under which this property holds, such as stationariness and large block size, are not satisfied. More on the use of the DCT in picture coding is presented in Chapter 12.

For smaller block lengths the use of suboptimal transforms may also help. Suppose that $\mathbf{M}$ does not diagonalize $\mathbf{C}$, but that the correlation matrix of $\mathbf{y}$ is given by

$$\mathcal{E}\{\mathbf{y}\mathbf{y}^T\} = \mathbf{M}\mathcal{E}\{\mathbf{x}\mathbf{x}^T\}\mathbf{M}^T = \mathbf{MCM}^T = \mathbf{C}'.$$

The variance of y_i is given by $c'_{i,i}$. Since $\mathbf{M}$ is orthogonal we obtain

$$\sigma_x^2 = C_{i,i} = \frac{1}{N}\sum_{j=1}^{N} C'_{j,j}, \quad i = 1, \ldots, N. \tag{5.63}$$

The average entropy per quantized transform coefficient is now given by

$$\begin{aligned}
\frac{1}{N}\sum_{i=1}^{N} H(\hat{y}_i) &\simeq \frac{1}{2}\log_2(2\pi e) + \frac{1}{N}\sum_{i=1}^{N} \log_2\left(\frac{\sqrt{C'_{i,i}}}{\Delta}\right) \\
&\leq \frac{1}{2}\log_2(2\pi e) + \log_2\left(\frac{\sqrt{\frac{1}{N}\sum_{j=1}^{N} C'_{j,j}}}{\Delta}\right) \\
&= \frac{1}{2}\log_2(2\pi e) + \log_2\left(\frac{\sigma_x}{\Delta}\right) \\
&= H(\hat{x}[n]). \tag{5.64}
\end{aligned}$$

The above expression only holds with equality for identical $C'_{i,i}$. The average coder complexity required to code quantizer-output symbols independently after transformation is now given by

$$\frac{K}{N}\sum_{i=1}^{N} \frac{\sqrt{C'_{i,i}}}{\Delta}.$$

It can be shown in this case that

$$\frac{K}{N}\sum_{i=1}^{N} \frac{\sqrt{C'_{i,i}}}{\Delta} \leq K\frac{\sigma_x}{\Delta}.$$

The above expression likewise only holds with equality for identical $C'_{i,i}$.

From the above it can be concluded that in this suboptimal case the average entropy of the quantized transform coefficients is less than or equal to the entropy of quantized input samples. A suboptimal orthogonal transform is therefore still capable of reducing the bit rate if it distributes the signal power non-uniformly over the y_i. An orthogonal transform that does this is said to concentrate the signal power. This is optimally achieved with the Karhunen-Loève transform.

The signal can be coded at the reduced bit rate with a coder complexity that equals the complexity required to code the input samples independently at a much higher bit rate.

It has been assumed that the signals are stationary and Gaussian. If the signal is not Gaussian, the Karhunen-Loève transform does not make the symbols $\hat{y}[n]$ independent, but merely uncorrelated. This means that after the transform the minimum bit rate cannot be reached by independent coding of the $\hat{y}[n]$. However, the transform may have the power-concentration property, and therefore still help to reduce the bit rate.

5.4.4 When the requirements are not satisfied

The above examples and derivations show how using an orthogonal transform can be helpful in source coding. Practical situations are often more complicated. For instance, the assumptions that $\Delta \ll \sigma_x$ or $\Delta \ll \sqrt{\lambda_i}$ are incorrect at low bit rates. Also, non-uniform quantizers may for perceptual reasons be used instead of uniform ones. In those cases the above derivation is incorrect. However, the general conclusion that the use of an orthogonal transform makes the coding easier remains valid in many cases.

A little more can be said about the case that the requirement $\Delta \ll \lambda_i$ is not satisfied for some i. The rate-distortion theory gives a more general answer to this problem [6]. Assume that $\sqrt{\lambda_i} \ll \Delta$ for some i. In that case, quantization would map those y_i on to 0. If it is known in advance for which i $\sqrt{\lambda_i} \ll \Delta$, no bits have to be spent to code these y_i. Assume that the number of y_i for which $\sqrt{\lambda_i} \ll \Delta$ is N_0, then the mean-square error is given by

$$D_{\text{MSE}} = \frac{N - N_0}{N} \frac{\Delta^2}{12} + \frac{1}{N} \sum_{i=N-N_0+1}^{N} \lambda_i. \tag{5.65}$$

If the λ_i, $i = N - N_0 + 1, \ldots, N$, are small enough, then

$$D_{\text{MSE}} < \frac{\Delta^2}{12}.$$

This implies that a greater step size Δ can be used than that prescribed by

$$D_{\text{MSE}} = \frac{\Delta^2}{12},$$

which leads to a further reduced bit rate. In practical applications a substantial part of the bit-rate reduction is often obtained by setting transform coefficients equal to zero. This is discussed further in Chapter 12.

5.4.5 Transform coding and perception

A transform can be regarded as a decomposition into basis vectors. This is easily demonstrated as follows. If

$$\mathbf{y} = \mathbf{M}\mathbf{x},$$

then, if $\mathbf{M}$ is orthogonal

$$\mathbf{x} = \mathbf{M}^T \mathbf{y},$$

or if $\mathbf{M}^T = (\mathbf{b}_1, \ldots, \mathbf{b}_N)$,

$$\mathbf{x} = \sum_{i=1}^{N} y_i \mathbf{b}_i.$$

Thus **x** is written as a sum of basis vectors. Similarly, in the two-dimensional case we have

$$\mathbf{X} = \sum_{k=1}^{N_v} \sum_{l=1}^{N_h} y_{k,l} \mathbf{B}_{k,l},$$

with

$$(\mathbf{B}_{k,l})_{i,j} = (\mathbf{M}_v{}^T)_{i,k} (\mathbf{M}_h)_{l,j}.$$

In both cases the input samples are decomposed as a weighted sum of basis components. Pictures of these basis components for an 8×8 two-dimensional discrete cosine transform can be found in Chapter 12.

There are some transforms of which the human observer is not able to perceive all basis components. The coefficients corresponding to these can be set equal to zero without introducing any perceptual distortion. Furthermore, for other components the human observer sometimes has a reduced sensitivity to errors, which means that their coefficients can be quantized less accurately. In this way the bit rate can be further reduced.

In what follows, the assumption that **M** is orthogonal is temporarily dropped. Assume that a perceptually desirable quality is obtained by quantizing the coefficients y_i, $i = 1, \ldots, N$, with quantization steps Δ_i, $i = 1, \ldots, N$, respectively. According to (5.39) and (5.38), D_{MSE} and the coding-error correlation matrix $\mathbf{C}_e$ are given by

$$D_{\text{MSE}} = \frac{1}{N} \sum_{i=1}^{N} \frac{\Delta_i^2}{12}, \tag{5.66}$$

and

$$\mathbf{C}_e = \mathbf{M}^{-1} \begin{pmatrix} \frac{\Delta_1^2}{12} & & \\ & \ddots & \\ & & \frac{\Delta_N^2}{12} \end{pmatrix} \mathbf{M}. \tag{5.67}$$

The matrix $\mathbf{C}_e$ in (5.67) is generally not diagonal, which means that the coding errors are correlated. If **M** is a discrete Fourier transform or a discrete cosine transform, then, for N large enough, it can be shown that the power spectral density of the coding error is approximated by

$$E(k\frac{2\pi}{N}) \simeq \frac{\Delta_k^2}{12}. \tag{5.68}$$

The error power spectral density at the decoder output is only flat if all step sizes are equal, otherwise it is not. The result of quantization with unequal step sizes is spectral shaping of the coding error and therefore, it can be seen as another form of noise-shaping.

For Gaussian $\mathbf{x}$ with correlation matrix $\mathbf{C}$, the correlation matrix of $\mathbf{y}$ is $\mathbf{MCM}^{-1}$. It is assumed that

$$\Delta_i \ll \sqrt{(\mathbf{MCM}^{-1})_{ii}}.$$

A simple extension of (5.14) shows that the minimum bit rate needed to code the quantized $\mathbf{y}$ is given by

$$\frac{1}{N}H(\hat{y}_1,\ldots,\hat{y}_N) \tag{5.69}$$

$$\simeq \frac{1}{2}\log_2(2\pi e) + \frac{1}{N}\log_2(\sqrt{|\mathbf{MCM}^{-1}|}) - \frac{1}{N}\sum_{i=1}^{N}\log_2(\Delta_i)$$

$$= \frac{1}{2}\log_2(2\pi e) + \frac{1}{N}\log_2(\sqrt{|\mathbf{C}|}) - \frac{1}{N}\sum_{i=1}^{N}\log_2(\Delta_i).$$

According to (5.51),

$$\frac{1}{N}H(\hat{y}_1,\ldots,\hat{y}_N) \tag{5.70}$$

$$\simeq \frac{1}{2}\log_2(2\pi e) + \frac{1}{N}\sum_{i=1}^{N}\log_2(\frac{\sqrt{\lambda_i}}{\Delta}) + \frac{1}{N}\sum_{i=1}^{N}\log_2(\frac{\Delta}{\Delta_i}).$$

The first two terms in the right-hand side of (5.70) give the entropy for a fixed step size Δ. The last term reflects the change if different step sizes are used. This shows how perceptual coding influences the bit rate. A good choice of step sizes yields

$$\frac{1}{N}\sum_{i=1}^{N}\log_2(\frac{\Delta}{\Delta_i}) < 0,$$

with imperceptible distortion. This implies a reduction of bit rate. Obviously,

$$\frac{1}{N}\sum_{i=1}^{N}\log_2(\frac{\Delta}{\Delta_i}) < 0,$$

if

$$\Delta_i \geq \Delta,\ i=1,\ldots,N.$$

If $\mathbf{M}$ does not diagonalize $\mathbf{C}$, as was assumed in Subsection 5.4.2, complicated dependent coding is required in order to achieve a bit rate close to (5.70). Therefore, it is often useful to specify the desired noise-shaping in the frequency domain and use a discrete Fourier transform or a discrete cosine transform because these transforms diagonalize $\mathbf{C}$ for N large enough. They consequently make noise-shaping possible without the necessity of dependent coding.

The above remark does not imply that other perceptually relevant transforms may not be useful. Even if the minimum bit rate is not achieved because optimal coding of quantizer-output symbols is too complicated, noise-shaping combined with independent coding may give a lower bit rate and better quality than can be obtained with a diagonalizing transform, a fixed step size Δ and independent coding. The reason is that the perceptually relevant transform may allow most of the Δ_i to be substantially larger than Δ. According to (5.70) and the power-concentration property (5.64), this may still yield a lower bit rate.

An example of the use of the discrete Fourier transform in this way is given in [53]. In that example a digital audio signal is split up into blocks of 1024 samples which are transformed to the frequency domain by a discrete Fourier transform. For each transform component it is computed whether it can be discarded or not. If it cannot be discarded, it is quantized with such an accuracy that the coding error cannot be perceived. It is claimed that a reduction in bit rate by a factor of seven or eight can be obtained in this way without introducing audible errors.

The same procedure is used in picture coding. Blocks of, for instance, 8×8 samples are transformed to the discrete cosine domain. Some of the transformed coefficients can be set to zero because the corresponding basis components cannot be observed. Others can be quantized with a step size such that the coding errors remain invisible. The quantization step size and the type of quantizer are often determined experimentally. More about selecting quantizers and step sizes is presented in Chapter 12.

5.5 Subband filtering

5.5.1 Introduction

In subband filtering the signal $x[n]$ is split up by means of a filter bank into a number of frequency bands, called subbands. The sample rate in each frequency band is reduced to twice its bandwidth. In this way the total sample rate of the signal remains unchanged. A reduction in the sample rate is also called decimation. The filter bank that splits the signal is therefore called a decimating filter bank. The separate frequency bands can be merged together to form a replica of the original signal by another filter bank which increases the sample rate of each frequency band. This process is called interpolation. All frequency bands with increased sample rate are subsequently added together. The merging filter bank is called an interpolating filter bank. If the filters in the decimating and interpolating filter banks are designed carefully, the output signal of the interpolating filter bank is, apart from a delay, a perfect copy of the input signal $x[n]$. There are many possibilities for interpolating and decimating filter banks, such as quadrature mirror filter banks (QMF), conjugate quadrature filter banks (CQF) and filter banks based on polyphase filtering [54, 55]. Moreover, the subbands can all have the same bandwidth, in which case we speak of a uniform filter bank, or have different bandwidths, in which case we have a non-uniform filter bank. Chapter 13 gives a

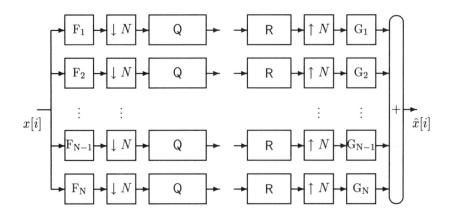

Figure 5.10 *Basic scheme of a subband source encoder and decoder.*

practical example of a QMF filter bank. We will regard the case of a uniform filter bank as shown in Figure 5.10. It should be remarked that non-uniform filter banks can also be brought into this form.

Assume that the subbands are numbered from 1 to N and that the transfer functions of the decimating and the interpolating filters in the ith subband are denoted by $F_i(\exp(j\theta))$ and $G_i(\exp(j\theta))$, respectively. Although it is not a necessary requirement for good coding results, it will be assumed in this section that the subband filters are derived from ideal bandpass filters with

$$H_i(\exp(j\theta)) = \begin{cases} 1, & (i-1)\frac{\pi}{N} \leq \theta < i\frac{\pi}{N}, \\ 0, & \text{otherwise.} \end{cases} \quad (5.71)$$

In that case

$$F_i(\exp(j\theta)) = H_i(\exp(j\theta)), \quad (5.72)$$

and

$$G_i(\exp(j\theta)) = N H_i(\exp(j\theta)). \quad (5.73)$$

Subband filtering is also used as a transformation for two-dimensional signals, such as pictures. This is extensively discussed in Chapter 13 and in [56]. The most common approach in the case of pictures is for the signal to be first filtered and decimated in one direction, e.g. the horizontal direction. The resulting subband signals are then filtered and decimated in the other direction.

The subband signals in the source decoder are quantized and coded. In the source decoder they are decoded, reconstructed and combined into a replica of the original. This is illustrated in Figure 5.10, which does not show the coder and decoder blocks.

The two reasons for applying subband filtering before quantizing and coding are the same as in the case of transform coding. The first is that it makes it easier to achieve a bit rate close to the signal's entropy. This can be illustrated for stationary Gaussian signals on the assumption that the mean-square error is the distortion measure. The second reason for employing subband filtering is that the amount of quantization noise in each subband can be easily controlled in such a way that it is perceptually least disturbing.

5.5.2 Subband coding and bit rate

Assume that the input signal $x[n]$ is zero-mean Gaussian and has a variance σ_x^2. It is split into N equally wide subbands. The subband signals are also Gaussian. Let σ_i^2 be the variance of the signal in subband i. It can be shown that

$$\sum_{i=1}^{N} \sigma_i^2 = \sigma_x^2.$$

If a quantizer with a step size $\Delta \ll \min(\sigma_1, \ldots, \sigma_N)$ operates on $x[n]$, the entropy of the quantized signal $\hat{x}[n]$ is given by

$$\begin{aligned} H(\hat{x}[n]) &\simeq \frac{1}{2}\log_2(2\pi e) + \log_2\left(\frac{\sigma_x}{\Delta}\right) \\ &= \frac{1}{2}\log_2(2\pi e) + \log_2\left(\frac{\sqrt{\sum_{i=1}^{N}\sigma_i^2}}{\Delta}\right). \end{aligned} \quad (5.74)$$

The distortion is given by

$$D_{\mathrm{MSE}} = \frac{\Delta^2}{12}.$$

It can be shown that distortion remains unchanged if the subband signals are quantized with step size $\frac{\Delta}{\sqrt{N}}$. The entropy of the quantized $y_i[n]$ in subband i is given by

$$H(\hat{y}_i[n]) \simeq \frac{1}{2}\log_2(2\pi e) + \log_2\left(\frac{\sqrt{N}\sigma_i}{\Delta}\right), \quad (5.75)$$

and the average entropy per quantized subband sample by

$$\frac{1}{N}\sum_{i=1}^{N} H(\hat{y}_i[n]) \simeq \frac{1}{2}\log_2(2\pi e) + \frac{1}{N}\sum_{i=1}^{N}\log_2\left(\frac{\sqrt{N}\sigma_i}{\Delta}\right). \quad (5.76)$$

If the σ_i^2 are replaced by $\frac{1}{N}\lambda_i$, expressions (5.74) and (5.76) are identical to (5.13) and (5.15), respectively. It therefore follows that in the case of subband coding

$$\frac{1}{N}\sum_{i=1}^{N} H(\hat{y}_i[m]) \leq H(\hat{x}[n]). \quad (5.77)$$

116 Signals and transformations

The above expression only holds with equality if $\sigma_1 = \sigma_2 = \ldots = \sigma_N$. A conclusion similar to that in the case of transform coding can be drawn here, namely that for the same distortion the average entropy of the quantized subband signals is less than the entropy of the quantized input signal, if subband filtering distributes signal power non-uniformly over the subbands. This is the subband-coding equivalence of the power-concentration property described in Subsection 5.4.3

Independent coding of the quantized $y_i[n]$ leads to a bit rate

$$\frac{1}{N} \sum_{i=1}^{N} H(\hat{y}_i[n]) \leq R < \frac{1}{N} \sum_{i=1}^{N} H(\hat{y}_i[n]) + 1 \tag{5.78}$$

Of course, the quantized $y_i[n]$ require different codes. If the subband filters are ideal bandpass filters, it can be shown that the subband signals are uncorrelated, or

$$\mathcal{E}\{y_i[m]y_j[n]\} = 0, \ i \neq j.$$

Because the $y_1[m], \ldots, y_N[m]$ are Gaussian, this means that

$$H(\hat{y}_1[m], \ldots, \hat{y}_N[m]) = \sum_{i=1}^{N} H(y_i[m]).$$

Combined coding of the quantized $y_1[m], \ldots, y_N[m]$ will therefore only further reduce the upper bound to the bit rate, which then satisfies

$$\frac{1}{N} \sum_{i=1}^{N} H(\hat{y}_i[n]) \leq R < \frac{1}{N} \sum_{i=1}^{N} H(\hat{y}_i[n]) + \frac{1}{N}. \tag{5.79}$$

By combined coding of groups of samples in each subband the upper bound can be still further reduced.

It is interesting to evaluate the performance of subband coding if the number of subbands increases. For that the following limit must be computed:

$$\lim_{N \to \infty} \frac{1}{N} \sum_{i=1}^{N} H(\hat{y}_i[m]) = \lim_{N \to \infty} \frac{1}{N} \sum_{i=1}^{N} \log_2 \left(\frac{\sqrt{N}\sigma_i}{\Delta} \right). \tag{5.80}$$

First, if $S_x(\theta)$ is the signal's spectral density, then

$$\sigma_x^2 = \frac{1}{2\pi} \int_{-\pi}^{\pi} S_x(\theta) \, d\theta = \frac{1}{\pi} \int_{0}^{\pi} S_x(\theta) \, d\theta, \tag{5.81}$$

and

$$\sigma_i^2 = \frac{1}{\pi} \int_{(i-1)\pi/N}^{i\pi/N} S_x(\theta) \, d\theta. \tag{5.82}$$

For large N we find that

$$\lim_{N \to \infty} \sigma_i^2 = \frac{1}{N} S_x \left(\frac{(2i-1)\pi}{2N} \right). \tag{5.83}$$

On substitution of (5.83) in (5.80) it follows that

$$\lim_{N \to \infty} \frac{1}{N} \sum_{i=1}^{N} H(\hat{y}_i[m]) \qquad (5.84)$$

$$= \frac{1}{2} \log_2(2\pi e) + \lim_{N \to \infty} \frac{1}{N} \sum_{i=1}^{N} \frac{1}{2} \log_2\left(S_x\left(\frac{(2i-1)\pi}{2N}\right)\right) - \log_2(\Delta)$$

$$= \frac{1}{2} \log_2(2\pi e) + \frac{1}{\pi} \int_0^\pi \frac{1}{2} \log_2(S_x(\theta)) \, d\theta - \log_2(\Delta)$$

$$= H_\infty(\hat{x}[n]),$$

which is identical to the result of (5.16). This means that a subband coder is capable of reaching the minimum bit rate (5.16) if the number of subbands is large enough.

The coder complexity of a subband coder is analysed as follows. Assume, as in the case of transform coding (5.54), that overload is negligible if the number of quantization levels is given by

$$M = K \frac{\sigma_x}{\Delta}$$

for the $x[n]$ and by

$$M_i = K \frac{\sqrt{N} \sigma_i}{\Delta} \qquad (5.85)$$

for the $y_i[n]$. For Gaussian signals, overload is negligible for $K = 8$.

Independent coding of the quantized $y_i[n]$ requires N Huffman codes with

$$\sum_{i=1}^{N} M_i = K \sum_{i=1}^{N} \frac{\sqrt{N} \sigma_i}{\Delta} \qquad (5.86)$$

entries in all. In this case it can be shown that the total average number of entries satisfies

$$\frac{1}{N} \sum_{i=1}^{N} M_i \leq K \frac{\sigma_x}{\Delta} \qquad (5.87)$$

and that (5.87) only holds with equality if the y_i have identical variances. Note that the average complexity is smaller than or equal to the complexity of independent Huffman coding of quantized input samples $x[n]$, but that, according to (5.77), the bit rate has been reduced. For N large enough, the bit rate approximates the bit rate that can be achieved by combined Huffman coding of N untransformed quantized samples.

It seems that the effects of subband filtering before quantizing are largely the same as the effects of an orthogonal transform. And that, for large N, the same performance and the same coder complexity can be expected.

5.5.3 Subband coding and perception

A subband coder can be perceptually optimized by not transmitting those subbands which cannot be observed and by quantizing others with step sizes
$$\frac{\Delta_i}{\sqrt{N}}.$$
It is assumed that
$$\Delta_i \ll \sigma_i.$$
As a result the power spectral density of the coding error in the output samples $\hat{x}[n]$ is given by
$$E(\theta) = \sum_{i=1}^{N} \frac{\Delta_i^2}{12} |H_i(\exp(\mathrm{j}\theta))|^2, \tag{5.88}$$
with $H_i(\exp(\mathrm{j}\theta))$ defined in (5.71). With proper choice of the step sizes, a stepwise approximation can be made of a desired error power spectral density. This desired error power spectral density should be chosen in such a way that the coding errors are imperceptible. The entropy, and thus the bit rate, follows from
$$\frac{1}{N} \sum_{i=1}^{N} H(\hat{y}_i[n]) \tag{5.89}$$
$$\simeq \frac{1}{2} \log_2(2\pi e) + \frac{1}{N} \sum_{i=1}^{N} \log_2 \left(\frac{\sqrt{N}\sigma_i}{\Delta_i} \right).$$
This can be compared with the entropy obtained with a fixed step size $\Delta/\sqrt{N}$. It follows that
$$\frac{1}{N} \sum_{i=1}^{N} H(\hat{y}_i[n]) \tag{5.90}$$
$$\simeq \frac{1}{2} \log_2(2\pi e) + \frac{1}{N} \sum_{i=1}^{N} \log_2 \left(\frac{\sqrt{N}\sigma_i}{\Delta} \right) + \frac{1}{N} \sum_{i=1}^{N} \log_2(\frac{\Delta}{\Delta_i}).$$
This result is similar to (5.70), derived for transform coding. The first two terms in the right-hand side of (5.90) give the entropy for a fixed step size Δ. The last term reflects the change if different step sizes are used. This shows how perceptual coding influences the bit rate. A good choice of step sizes yields
$$\frac{1}{N} \sum_{i=1}^{N} \log_2(\frac{\Delta}{\Delta_i}) < 0,$$
with imperceptible distortion. This implies a reduction of bit rate. Obviously
$$\frac{1}{N} \sum_{i=1}^{N} \log_2(\frac{\Delta}{\Delta_i}) < 0,$$
if
$$\Delta_i \geq \Delta, \ i = 1, \ldots, N.$$

5.6 Predictive filtering

5.6.1 Introduction

In a predictive coder the signal is filtered before quantization and coding. After decoding and reconstruction it is filtered with the inverse of the filter in the source encoder. The output of the filter in the source encoder is the difference between a prediction of $x[n]$ and $x[n]$ itself. The prediction is computed with a so-called predictive filter. This is usually a non-recursive filter that predicts a sample $x[n]$ as a linear combination of p previous samples. The integer p is called the order of prediction. The difference between the prediction of $x[n]$ and $x[n]$ is called the prediction error. It is quantized and coded. Source coding with predictive filtering is called linear predictive coding (LPC). Linear predictive coding is often used in speech coding [57].

In this section it is shown, on the assumptions that the signal $x[n]$ is Gaussian and stationary and that the coding-error variance is the distortion measure, that predictive filtering helps to reduce the bit rate. This is done in Subsection 5.6.2. Quantization in a predictive coder is less straightforward than in a transform or subband coder. To obtain a minimum quantization-error variance a noise-shaping quantizer must be used. It is shown in Subsection 5.6.3 that the above approach results in the well-known differential pulse-code modulation (DPCM) [4] system. A perceptual spectral weighting of the quantization errors can be achieved by adapting the noise-shaping quantizer.

5.6.2 Predictive filtering and bit rate

In linear predictive source coders the prediction of $x[n]$ is obtained by a linear combination of $x[n-p], \ldots, x[n-1]$. The prediction can be seen as the output of a finite impulse response filter. The prediction error is the difference between $x[n]$ and its prediction. Of course, the prediction error can also be seen as the output of a finite impulse response filter whose input sequence is $x[n]$. In this section a somewhat more general approach is followed, as we allow the filter that computes the prediction error to have an infinite impulse response.

It has been shown in Section 5.2 that for Gaussian signals quantized with step size Δ (5.16) and (5.17) and (5.20) express the relation between the power spectral density and the entropy. The mean-square error after decoding and reconstruction is given by

$$D_{\text{MSE}} = \frac{\Delta^2}{12}.$$

Filtering a signal changes its power spectral density. The question arises whether a filtering operation that converts a signal $x[n]$ into a white-noise signal, which after quantization with step size Δ has an entropy $H_\infty(\hat{x}[n])$ and can be coded

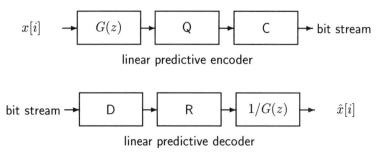

Figure 5.11 *Blocks in a linear predictive source-coding system.*

easily, is a suitable transformation. After decoding and reconstruction, the source decoder computes a replica of the input signal by an inverse filtering operation. This approach is analysed below. The basic coding scheme is shown in Figure 5.11.

Filtering the signal $x[n]$ with a filter with a transfer function

$$G(z) = \frac{A(z)}{B(z)} = \frac{1 + a_1 z^{-1} + \ldots + a_p z^{-p}}{1 + b_1 z^{-1} + \ldots + b_p z^{-p}}$$

results in a Gaussian white-noise signal $y[n]$ with a variance σ_ϵ^2 which, after quantization with a step size Δ, has an entropy

$$H(\hat{y}[n]) \simeq \frac{1}{2} \log_2(2\pi e) + \log_2\left(\frac{\sigma_\epsilon}{\Delta}\right).$$

This entropy is identical to $H_\infty(\hat{x}[n])$. However, if this signal is quantized with a step size Δ, after decoding, reconstruction and inverse filtering with a transfer function

$$\frac{B(\exp(j\theta))}{A(\exp(j\theta))}$$

a coding error results with a power spectral density

$$S_{dd}(\theta) = \frac{\Delta^2}{12} \frac{|B(\exp(j\theta))|^2}{|A(\exp(j\theta))|^2}. \tag{5.91}$$

The distortion in the source decoders output is given by

$$D_{\text{MSE}} = \frac{1}{2\pi} \int_{-\pi}^{\pi} S_{dd}(\theta) \, d\theta \geq \frac{\Delta^2}{12}.$$

The above approach leads to the desired bit rate, but the distortion is greater than what would have been obtained by quantizing $x[n]$ with step size Δ.

The distortion can be reduced by using a noise-shaping quantizer such as has been described in Section 4.4.4 and Figure 4.11 on page 68. If the noise-shaping filter in this figure has a transfer function

$$F(z) = \frac{A(z)}{B(z)} - 1$$

and an impulse response

$$1, f_1, f_2, \ldots,$$

it follows from (4.18) and (5.91) that

$$S_{dd}(\theta) = |1 + F(\exp(j\theta))|^2 \frac{|B(\exp(j\theta))|^2}{|A(\exp(j\theta))|^2} \frac{\Delta^2}{12}$$

$$= \frac{\Delta^2}{12}. \tag{5.92}$$

The entropy of the quantized $y[n]$ has increased slightly because the sum of the $y[n]$ and the output of the noise-shaping filter is quantized. On the basis of the central limit theorem [10] it can be assumed that the output of the noise-shaping filter is Gaussian. This means that the entropy of the quantizer-output symbols is given by

$$H(\hat{y}[n]) \tag{5.93}$$

$$\simeq \frac{1}{2} \log_2(2\pi e) + \log_2 \left(\frac{\sqrt{\sigma_\epsilon^2 + \frac{\Delta^2}{12} \sum_{i=1}^{\infty} f_i^2}}{\Delta} \right)$$

$$= \frac{1}{2} \log_2(2\pi e) + \log_2 \left(\frac{\sigma_\epsilon}{\Delta} \right) + \log_2 \left(\sqrt{1 + \frac{\Delta^2}{12\sigma_\epsilon^2} \sum_{i=1}^{\infty} f_i^2} \right).$$

If $\Delta \ll \sigma_\epsilon$, then $H(\hat{y}[n]) \simeq H_\infty(\hat{x}[m])$. The number of entries for a Huffman table to code the quantizer-output symbols is given by

$$M = K \frac{\sigma_\epsilon}{\Delta} \left(\sqrt{1 + \frac{\Delta^2}{12\sigma_\epsilon^2} \sum_{i=1}^{\infty} f_i^2} \right). \tag{5.94}$$

If $\Delta \ll \sigma_\epsilon$, then

$$M \simeq K \frac{\sigma_\epsilon}{\Delta} \leq K \frac{\sigma_x}{\Delta}, \tag{5.95}$$

where σ_x is the square root of the input signal's variance. Expression (5.95) holds with equality if and only if $x[n]$ is white noise. This shows that with predictive filtering the signal can be coded at a bit rate R satisfying

$$H_\infty(\hat{x}[n]) \leq R < H_\infty(\hat{x}[n]) + 1, \tag{5.96}$$

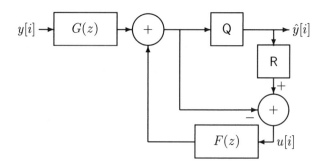

Figure 5.12 *Source encoder of a linear predictive coding system. The coder has been omitted.*

and a distortion

$$D_{\text{MSE}} = \frac{\Delta^2}{12},$$

while the coder complexity is less than that required for independent coding of quantized $x[n]$.

The ratio

$$\frac{\sigma_x^2}{\sigma_\epsilon^2}$$

is generally referred to as the prediction gain. It is usually expressed in dB. Higher prediction gains result in a greater difference between $H_\infty(\hat{x}[n])$ and $H(\hat{x}[n])$ and a greater reduction in coder complexity.

Figure 5.12 shows the source encoder of a linear predictive coder. The coder is omitted.

5.6.3 Linear prediction

We will now turn to what is generally known as linear prediction: the case of a non-recursive predictive filter. In the most common form of a linear predictive coder we have $B(z) = 1$. The filter in the source encoder has then a transfer function which is often written as

$$G(z) = 1 + a_1 z^{-1} + \ldots + a_p z^{-p} = 1 - A'(z).$$

Figure 5.13 shows possible implementations of the filters in the source encoder and decoder and Figure 5.14 shows an example of a straightforward realization of the filter with a third-order transfer function $A'(z)$.

As already stated, a linear prediction of the sample $x[n]$ based on p previous samples is computed. This explains the name predictive filtering. If the prediction

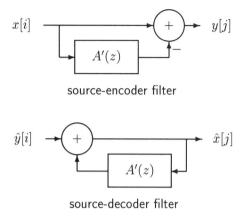

Figure 5.13 *Filters of a non-recursive linear predictive coding system.*

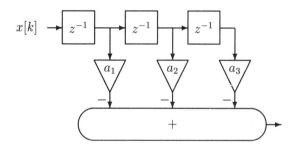

Figure 5.14 *Example of a third-order filter $A'(z)$.*

is good the prediction gain is high and the signal $y[n]$ has a small variance. In that case the entropy of the quantized $y[n]$ is also small. Other filter structures, such as lattice filters, are also used to implement the source encoder and decoder filters. An overview is given in [57].

A source encoder without a coding block is illustrated in Figure 5.15. It can be shown by some manipulations with the diagram that this encoder is identical to that shown in Figure 5.16, which is generally known as the DPCM encoder [4]. DPCM is discussed further in Chapter 11.

Note that in order to capture the DPCM encoder of Figure 5.16 in the simple 'TQC' model of Figure 2.6 on page 24, the latter model has to be extended to the

124 Signals and transformations

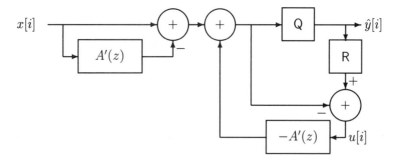

Figure 5.15 *Source encoder of a non-recursive linear predictive coding system. The coder has been omitted.*

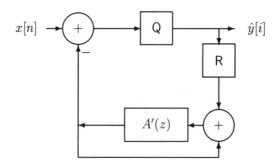

Figure 5.16 *DPCM encoder.*

one shown in Figure 5.17. The reconstructor and the filtering operation are in this view considered to be part of the transformation block in the encoder.

Linear predictive coding has certain advantages over transform and subband coding. First of all, only one code is needed for the quantized $y[n]$, whereas in the case of transform and subband coding N codes may be required. Secondly, predictive coders can be very simple, e.g. the DPCM coder of Example 1. Transform

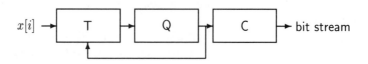

Figure 5.17 *Source encoder with feedback.*

and subband coders are usually more complicated. Thirdly, for some signals, such as speech and music the power spectral density can be modelled accurately as

$$\sigma_\epsilon^2 \frac{1}{|1 - A'(\exp(j\theta))|^2},$$

with a small order of prediction p, e.g. $p = 12$. Moreover, the prediction coefficients $a_1, \ldots, a_p$ can be estimated easily from the input signal, which means that this coding method can readily be made adaptive. In that case the source-coding system is sometimes called adaptive DPCM, or ADPCM.

Example 22 Recall that in Example 19 on page 94 $\Delta = 0.1$, $a_1 = -0.9$, $a_2 = 0.81$, $\sigma_x = 1$ and $\sigma_e = 0.508$. The encoder of Figure 5.15 is used to encode the signal. The entropy of the quantizer-output symbols is given by (5.93) and is

$$H(\hat{y}[n]) = 4.395,$$

which is just a fraction above

$$H_\infty(\hat{x}[n]) = 4.392.$$

The number of entries in the Huffman table, required to achieve a bit rate between 4.395 and 5.395 is given by (5.95). With $K = 8$ the result is

$$M = 41.$$

Coding larger groups of quantizer-output symbols reduces the upper bound to the bit rate at the expense of increased coder complexity. The lower bound is shown in Figure 5.4.

In the above example linear predictive coding outperforms transform and subband coding. This is because the predictive filter closely matches the signal model; it is not a general conclusion. Although linear predictive coding has the advantages of a relatively simple source encoder and that of being easily made adaptive, it also has some disadvantages. One is that, because of the recursive filter in the source decoder, the system is more sensitive to uncorrected channel errors. Especially in the adaptive case where filter coefficients are transmitted to the decoder, this filter can become unstable due to transmission or rounding errors. Perceptual predictive coding, as will be shown in Subsection 5.6.4, also has more limitations than perceptual subband and transform coding.

5.6.4 Predictive filtering and perception

Perceptual noise-shaping can be obtained by choosing the transfer function of the noise-shaping filter in Figure 5.12 as

$$F(z) = (\frac{A(z)}{B(z)} - 1) F'(z).$$

In that case the source-coding error has a power spectral density

$$S_{dd}(\theta) = \frac{\Delta^2}{12} |F'(\exp(j\theta))|^2 . \tag{5.97}$$

The desired noise-shaping can be obtained by an appropriate choice of $F'(z)$, but note should be taken of the remarks made on the risks involved in noise-shaping at the end of Section 4.4.4.

In subband and transform coding the possibility exists not to transmit certain subbands or transform coefficients. This is especially useful if those subbands or coefficients cannot be perceived. This possibility is not present in predictive coding, which is a disadvantage in comparison with the other two source-coding techniques.

5.6.5 Analysis-by-synthesis methods

There is an interesting class of source-coding systems for speech which are based on linear prediction and vector quantization, often in combination with perceptual weighting of the coding error. These systems are called codebook-excited linear predictive (CELP) coders. This subsection discusses an elementary CELP coder.

In a CELP-based speech coder, an excitation sequence or vector is selected from a codebook on the basis of a weighted mean-square error criterion. The excitation sequences typically contain 10–20 samples. The codebook contains about 1000 sequences. Selection is done by feeding all excitation sequences into a replica of the source decoder, which is similar to a normal ADPCM decoder. The output sequences thus generated are compared with the original input sequences. The criterion for comparison is the weighted mean-square error. This means that the difference between the input and generated output is passed through a noise-weighting filter. The power of the filtered difference sequence is then estimated. This power is called the weighted mean-square error. The excitation sequence yielding the minimum weighted mean-square error is selected and its index transmitted. This procedure is often called analysis-by-synthesis.

Figure 5.18 shows how a weighting filter is used in a CELP coder. From the incoming speech signal $x[i]$ the LPC analysis box computes the prediction coefficients $a_1, \ldots, a_p$, the coefficients for the weighting filter and a gain factor. The codebook contains a number of excitation vectors. The vector length is N. During selection all vectors are multiplied by the gain factor and passed through an analysis filter. This results in a sequence of N samples denoted by $\hat{x}[i]$. An error sequence is formed by subtracting N input samples $x[i]$ from N samples $\hat{x}[i]$. The error sequence is passed through the weighting filter. The weighted mean-square error, which is the short-term power of the weighted error sequence, is computed. The selector box selects the code vector that results in the lowest weighted mean-square error. The quantized and coded gain factor and prediction coefficients and the index of the vector are transmitted to the source decoder.

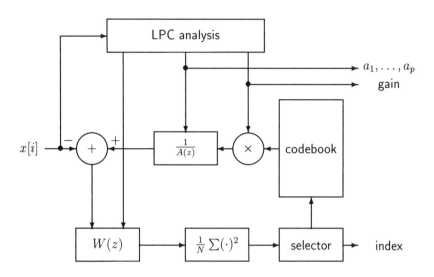

Figure 5.18 *Simplified CELP encoder without pitch prediction.*

The commonly used weighting filters are based on the prediction coefficients $a_1, \ldots, a_p$ of the speech signal. A possible form of a weighting filter is [58]

$$W(z) = \frac{1 + \sum_{k=1}^{p} a_k \gamma_1^k z^{-k}}{1 + \sum_{k=1}^{p} a_k \gamma_2^k z^{-k}} \frac{1}{1 + \sum_{k=1}^{2} p_k \delta^k z^{-k}}. \tag{5.98}$$

The coefficients p_1 and p_2 are found by applying LPC analysis to the first three autocorrelation lags of the sequence. The coefficients δ, γ_1 and γ_2 control the amount of weighting at the position of the formants. They are such that good perceptual performance is obtained. According to [58],

$$\delta = 0.7, \gamma_1 = 0.95, \gamma_2 = 0.8.$$

Other, similar, forms have also been reported.

For a well-chosen codebook that is large enough, the power spectral density function of the source-coding error is proportional to

$$\left| \frac{1}{W(\exp(j\theta))} \right|^2.$$

5.7 Problems

Problem 23 Use (5.12) and (5.60) to derive (5.61).

128   Signals and transformations

Problem 24 Give the transform matrix of Example 5 on page 19.

Problem 25 Give the equivalent for Figure 5.7 on page 99 for a subband coder. What is the noise-power spectral density if the quantizers in subbands $1, \ldots, N$ have step sizes Δ_i, $i = 1, \ldots, N$.

Problem 26 Give the equivalent for Figure 5.7 on page 99 for a predictive coder.

Problem 27 Show the equivalence between Figure 5.15 and Figure 5.16.

Problem 28 Draw a source decoder for the CELP encoder in Figure 5.18. Which type of vector quantization is used?

Problem 29 What is the function of the reconstructors in Figure 5.18?

6

Adaptive source coding

6.1 Introduction

In the preceding chapters we generally assumed that input signals were stationary, or at least second-order or wide-sense stationary [10]. Roughly speaking, these assumptions indicate that statistical properties such as the mean, the autocorrelation function, the entropy and the probability density function do not depend on time, or, in the case of pictures, on the spatial coordinates.

If a designer of a source-coding system observes that the signal parameters used in a coding system, e.g. the prediction coefficients, change more with time than can be expected on the basis of the assumption of stationariness, he may adopt a non-stationary signal model which allows these parameters to vary with time. It may be difficult to parametrize such a non-stationary model. It is then often assumed that the variations of the parameters are so slow that the parameters can be estimated or updated regularly and that the source encoder and decoder can be adapted to these parameters. This is called adaptive source coding.

This chapter concentrates on different forms of adaptive source-coding systems and their properties. Adaptivity may occur in all basic blocks of a source-coding system. For instance, if in Figure 5.15 on page 124 the prediction coefficients are computed adaptively, then the transformation block is adaptive. If a source-coding system switches between quantizers, the quantization blocks are adaptive. Ziv-Lempel coding, as discussed in Subsection 3.5.3, is an example of adaptive coding.

The further contents of this chapter are as follows. Section 6.2 discusses the two principal forms of adaptivity: forward and backward adaptivity. Sections 6.3, 6.4 and 6.5 show that adaptation occurs at a certain rate, requires parameter estimation and can influence coding delay. Section 6.6 gives a brief comparison of forward- and backward-adaptive source coders.

There are too many different forms of adaptive source-coding systems, and ways to make basic blocks of source-coding schemes adaptive, for them all to be dealt with here. Therefore, Sections 6.7, 6.8 and 6.9 will only give a few general remarks on adaptivity in the three types of source coding that have been discussed in Chapter 5: transform, subband and predictive coding. Section 6.10 discusses two forms of

130 Adaptive source coding

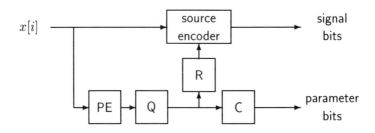

Figure 6.1 *Simplified diagram of a forward-adaptive source encoder.*

bit-rate control: adaptive bit allocation and buffer control.

6.2 Forward and backward adaptivity

6.2.1 Forward adaptivity

Assume that the source encoder and decoder depend on a set of signal parameters denoted by $\mathcal{P}$. The set $\mathcal{P}$ can, for instance, contain the peak value which is used in the APCM coder of Example 5 on page 19 or the prediction coefficients used in an adaptive linear predictive coder. In a forward-adaptive source coder the parameters in set $\mathcal{P}$ are estimated from the input signal of the source encoder. Figures 6.1 and 6.2 are principle diagrams of a forward-adaptive source encoder and decoder, respectively. It must be remarked that these principle diagrams are incomplete because they do not show some other important aspects of adaptivity which will be discussed in Sections 6.3, 6.4 and 6.5. The set of parameters is estimated by the box PE. After estimation the parameters are quantized, coded and transmitted to the source decoder. In addition, the quantized parameters are reconstructed and used in the source encoder. It is essential to quantize and reconstruct these parameters before using them because source encoder and decoder must use identical sets of parameters.

6.2.2 Backward adaptivity

In a backward-adaptive source coder the parameters for adaptation are estimated from previous output samples of the decoder. For this, a replica of the source decoder must be present in the source encoder. Figures 6.3 and 6.4 show principle diagrams of a backward-adaptive source encoder and decoder. The delay after parameter estimation ensures that parameters estimated from past decoder-output

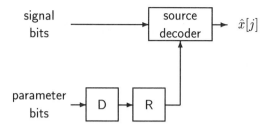

Figure 6.2 *Simplified diagram of a forward-adaptive source decoder.*

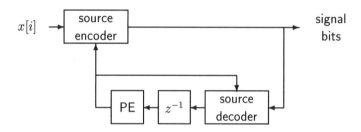

Figure 6.3 *Simplified diagram of a backward-adaptive source encoder.*

samples are used. As in the case of forward-adaptive source coders, it must be remarked that this principle diagram is also incomplete. The set of parameters is estimated by the box PE. Note that after estimation they do not have to be quantized since they are not transmitted.

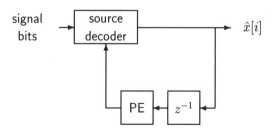

Figure 6.4 *Simplified diagram of a backward-adaptive source decoder.*

6.2.3 Delay

In the following sections it is sometimes necessary to take into account the delays which occur in the source encoder, source decoder and parameter estimation boxes of Figures 6.1 to 6.4. All delays are expressed in input sample periods. The parameter-estimation delay is denoted by L_P, the source-encoder delay by L_E, and the source-decoder delay by L_D. In Figures 6.1 to 6.4 the delays can be thought to be at the inputs as well as at the outputs of the boxes PE, source encoder and source decoder, respectively. Delays occur, for instance, if the boxes contain delay-introducing filter operations, as in the case of the subband coder discussed in Section 5.5.

Only so-called algorithmic delays are considered here. In hardware implementations an implementation delay is sometimes introduced, for instance, because intermediate results are doubly buffered.

6.3 Adaptation rate

Figures 6.1 to 6.4 suggest that a new set of parameters is computed for every time index k. In the case of forward-adaptive source-coding systems this would increase the bit rate for parameter transmission excessively. This is not true for backward-adaptive source-coding systems, but both in forward- and backward-adaptive source-coding systems the complexity would become unrealistically high. Usually the adaptation rate is a fraction $1/K$ of the sample rate. The reciprocal K of the adaptation rate will be called the adaptation period. Figures 6.5, 6.6 and 6.7 show basic diagrams of forward- and backward-adaptive source coders with an adaptation rate $1/K$.

Example 23 Consider the APCM coder of Example 5. Clearly

$$\mathcal{P} = \{\hat{x}_{\max}\}.$$

As the $\hat{x}_{\max}$ is derived from the input samples, it follows that the APCM source coder of this example is forward-adaptive. The parameter $\hat{x}_{\max}$ is computed every N samples, therefore the adaptation period K is equal to the block length N.
◇

The relation $N = K$, observed in Example 23, is often valid for block-based source coders such as transform coders and APCM.

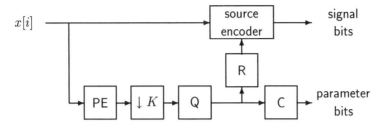

Figure 6.5 *Simplified diagram of a forward-adaptive source encoder with adaptation rate $1/K$.*

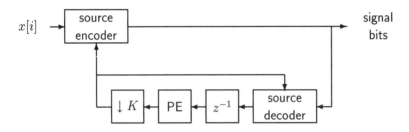

Figure 6.6 *Simplified diagram of a backward-adaptive source encoder with adaptation rate $1/K$.*

6.4 Parameter estimation

The way parameters are estimated in adaptive source-coding systems largely depends on the kind of parameters. Therefore, only general remarks are made in the following paragraphs.

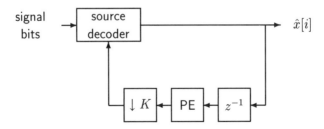

Figure 6.7 *Simplified diagram of a backward-adaptive source decoder with adaptation rate $1/K$.*

134 Adaptive source coding

In Figures 6.1 to 6.4 the parameters are estimated from the input samples or from the decoded output samples. In practical systems this may not be the case because it is sometimes more advantageous to estimate parameters from another point in the encoder or decoder, e.g. from the transformed input samples. However, this situation can always be translated to the situation that is depicted in Figures 6.1 to 6.4.

Two ways of parameter estimation can be distinguished: recursive and non-recursive. Non-recursive estimators estimate parameters from a finite sequence of input samples or decoded output samples. Recursive estimators use one or more previously estimated parameters in addition. Let $\mathcal{P}[k]$ denote the parameter set used from time index kK to time index $kK + K - 1$. Let Q denote the number of past input samples, in the case of forward-adaptive source coding, or decoded output samples, in the case of backward-adaptive source coding, that are involved in parameter estimation. Finally, let P denote the number of previously estimated parameters. Non-recursive forward-adaptive parameter estimation with adaptation rate $1/K$ is then described by

$$\mathcal{P}[l] = f(x[lK - Q + 1], \ldots, x[lK]), \tag{6.1}$$

and recursive forward-adaptive parameter estimation with adaptation rate $1/K$ by

$$\mathcal{P}[l] = f(x[lK - Q + 1], \ldots, x[lK], \mathcal{P}[l - P], \ldots, \mathcal{P}[l - 1]). \tag{6.2}$$

Backward-adaptive non-recursive parameter estimation with adaptation rate $1/K$ is described by

$$\mathcal{P}[l] = f(\hat{x}[lK - Q], \ldots, \hat{x}[lK - 1]), \tag{6.3}$$

and recursive backward-adaptive parameter estimation with adaptation rate $1/K$ by

$$\mathcal{P}[l] = f(\hat{x}[lK - Q], \ldots, \hat{x}[lK - 1], \mathcal{P}[l - P], \ldots, \mathcal{P}[l - 1]). \tag{6.4}$$

The following example shows a simple case of first-order recursive parameter estimation.

Example 24 Consider the APCM coder of Example 5 on page 19, but now assume that the quantizer has been optimized for Gaussian inputs with $\sigma^2 = 1$. In that case $\hat{x}_{\max}$ must be replaced by an estimate of σ. Usually a quantized version of

$$\sqrt{\frac{1}{N} \sum_{i=0}^{N-1} x_i^2}$$

is taken. If N is short, the estimate will not be reliable. The estimate can be made recursive by choosing

$$\mathcal{P}[l] = (1 - \alpha)\frac{1}{N} \sum_{i=0}^{N-1} x_i^2 + \alpha \mathcal{P}[l - 1].$$

Now $\hat{x}_{\max}$ must be replaced by

$$\sqrt{\mathcal{P}[l]}.$$

If α is close to 1, a sudden jump in

$$\sqrt{\frac{1}{N}\sum_{i=0}^{N-1} x_i^2}$$

will result in considerable overload distortion, as $\mathcal{P}[l]$ will not be able to track this jump. It will take some time before $\mathcal{P}[l]$ has adapted to this change.

◇

In the case of non-recursive parameter estimation the estimates are solely based on a finite number of input samples or decoded output samples. In the case of a recursive parameter estimation the estimates are based on all previous input samples or decoder output samples. Note that even in the case of non-recursive backward-adaptive parameter estimation the estimates are based on all previous input samples. Recursive estimation has the advantage that the same accuracy can be achieved with less input samples or decoded output samples.

Often Q is a integer multiple of K. This simplifies (6.1), (6.2), (6.3) and (6.4). If $Q \leq K$, we speak of parameter estimation with non-overlapping segments or non-overlapping blocks. The APCM coder of Example 5 on page 19, for instance, has non-overlapping blocks. If $Q > K$, we speak of overlapping segments or overlapping blocks. For the same K, non-overlapping blocks are usually computationally more efficient than overlapping blocks. In the case of overlapping blocks however, the parameter estimates can be more accurate since they are based on more data. On the other hand, the tracking of changes in the characteristics of the data becomes slower.

Example 25 Consider again the APCM coder of Example 5 on page 19 and assume as in Example 24 that the quantizer has been optimized for Gaussian inputs with $\sigma^2 = 1$. In a non-recursive version with overlapping blocks, σ is estimated by

$$\mathcal{P}[l] = \sqrt{\frac{1}{(qN)} \sum_{i=-(q-1)N}^{N-1} x_i^2},$$

Where q is the number of blocks involved in parameter estimation. If q is large, a sudden jump in

$$\sqrt{\frac{1}{N}\sum_{i=0}^{N-1} x_i^2}$$

will result in considerable overload distortion, as $\mathcal{P}[l]$ will not be able to track this jump. It will take some time before $\mathcal{P}[l]$ has adapted to this change.

◇

136 Adaptive source coding

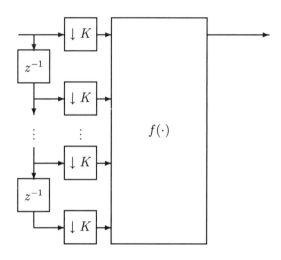

Figure 6.8 *Non-recursive parameter estimation with adaptation rate $1/K$. The input is either the sequence $x[k]$ or $\hat{x}[k]$. The output is the parameter set $\mathcal{P}[l]$.*

A realization according to Figures 6.5, 6.6 and 6.7 would be inefficient since parameters are computed for every time index k and subsequently decimated. Practically, parameters are only computed every K sampling periods. Figure 6.8 shows a more realistic approach to non-recursive parameter estimation in which parameters are estimated every Kth time index from an input sequence or decoded output sequence of length Q. Figure 6.9 shows the recursive equivalent.

Example 26 The blockwise scaling in Figure 2.2 on page 22 can be seen as a very simple adaptive transform. In that sense, Figure 2.2 shows an adaptive-transform-coder interpretation of an APCM encoder. The incorrect Figure 2.4 on page 23 seems to show an adaptive quantization interpretation of an APCM coder, since the combination of scaler and quantizer can be seen as an adaptive quantizer. The adaptation period of the quantizer is N samples. Figure 6.10 shows a correct version of Figure 2.4, with the proper delays introduced, and clearly showing from which input samples the scale factor is determined. The adaptive quantization interpretation seems to justify the name APCM better than the adaptive-transform-coder interpretation.
◇

So far, four types of adaptive source coders have been distinguished: forward-adaptive non-recursive, forward-adaptive recursive, backward-adaptive non-recursive, and backward-adaptive recursive ones. Only in the forward-adaptive non-recursive coder is the number of input samples that are used to estimate parameters finite. During parameter estimation they can be thought to be stored in a parameter-estimation memory containing Q samples. In the other cases the number of input samples involved in parameter estimation is infinite. However, the number of input samples which effectively contribute to parameter estimation is

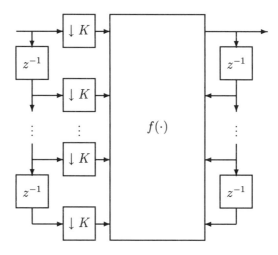

Figure 6.9 *Recursive parameter estimation with adaptation rate $1/K$. The input is either the sequence $x[k]$ or $\hat{x}[k]$. The output is the parameter set $\mathcal{P}[l]$.*

generally finite. It is called the effective parameter-estimation memory size and will be denoted by Q'. How large Q' is, largely depends on the estimation algorithm and on the source-coding method. Generally, for the same Q, the Q' of a recursive source-coding system will be larger than the Q' of the corresponding non-recursive source-coding system. In what follows both parameter-estimation memory size and effective parameter-estimation memory size are denoted by Q'.

The effective parameter-estimation memory size determines the time resolution of the adaptation. If a signal parameter varies within Q' samples, the variations can generally not be tracked.

6.5 Adaptation and coding delay

This section discusses adaptation and coding delay. Coding delay is more important for music and speech source-coding systems than for source-coding systems for two-dimensional signals such as pictures. The reason is that in the latter systems the 'time' indices i, j in $x[i, j]$ are associated with space and not with time, which makes the delay meaningless. Adaptation delay is meaningful in both one- and two-dimensional source-coding systems, as it refers to the index shift between the samples from which parameters are estimated and the samples to which the parameters are applied. Only one-dimensional signals are considered here.

Not only is the number Q' of input samples from which estimated parameters are derived important, but also their specific location in time. This location is

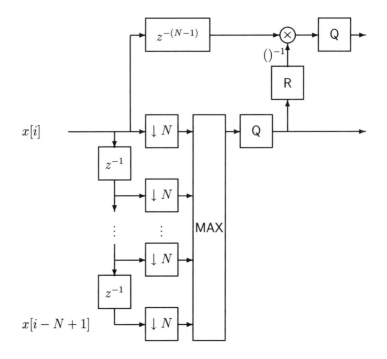

Figure 6.10 *APCM encoder with a block length N. The coder block has been omitted*

different for the four types of adaptive coders. It depends on the value Q', but also on the delays in source encoding, source decoding and parameter estimation.

For forward-adaptive coders the interval from which $\mathcal{P}[l]$ is derived is given by

$$lK - L_\mathrm{P} - Q' + 1 \leq k \leq lK - L_\mathrm{P}. \tag{6.5}$$

For backward-adaptive coders this interval is given by

$$lK - L_\mathrm{P} - L_\mathrm{E} - L_\mathrm{D} - Q' \leq k \leq lK - L_\mathrm{P} - L_\mathrm{E} - L_\mathrm{D} - 1. \tag{6.6}$$

The intervals to which the parameters are applied may be different from the intervals (6.5) and (6.6). Generally, the parameters are applied to the current adaptation interval, which is in all cases given by

$$lK - L_\mathrm{E} \leq k \leq (l+1)K - L_\mathrm{E} - 1 \tag{6.7}$$

It follows from (6.5), (6.6) and (6.7) that parameters are derived from a certain interval and may be applied to another. The interval (6.7) must be compared with intervals (6.5) and (6.6) to determine whether adaptation runs behind or ahead of the signal. For this purpose it is most convenient to associate time indices with the intervals and subtract those. The difference between the time index associated with the interval to which parameters are applied and the time index associated with the interval for which they are valid is called the adaptation delay. It quantifies the number of samples by which adaptation is running behind or ahead.

Suitable time indices are the upper edge or the centre of the interval. Although the choice is somewhat arbitrary here, the centre is chosen.

The adaptation delay of a forward-adaptive source coder is given by

$$L_{AF} = \frac{1}{2}(Q' + K) + L_P - L_E - 1. \qquad (6.8)$$

The adaptation delay of the forward-adaptive source coder can be positive or negative, which means that adaptation runs ahead of changes in signal characteristics. It can be made zero by inserting delay elements before the source encoder if L_{AF} is positive or by inserting delay elements before the parameter estimation block if L_{AF} is negative. A special case is the non-recursive forward-adaptive source coder with parameter estimation with non-overlapping blocks, i.e. $Q = K$, and equal encoder and estimation delay, i.e. $L_P = L_E$. In that case, an additional delay of $K - 1$ sample periods makes the adaptation delay equal to zero. One example is the forward-adaptive non-recursive transform coder, with $N = K = Q$, where N is the block length of the transform or the APCM coder of Example 5 on page 19. If L_P and L_E cannot be influenced, adaptation can be made faster by decreasing Q' or K. A small Q' has the disadvantage of an increased parameter estimation variance. A small K increases hardware complexity and bit rate, as more parameters have to be transmitted.

The adaptation delay of a backward-adaptive source coder is given by

$$L_{AB} = \frac{1}{2}(Q' + K) + L_P + L_D. \qquad (6.9)$$

Note that the delay in the source encoder does not influence the adaptation delay of the backward-adaptive source coder, but the delay of the source decoder does. As all terms in (6.9) are positive, the adaptation delay of a backward-adaptive source coder cannot be made zero. Adaptation is faster when Q' and K are small. A small Q' has the disadvantage of an increased parameter estimation variance. A small K increases hardware complexity, but has no influence on the bit rate.

Another aspect of source coders is the delay L_C between the input of the source encoder and the output of the source decoder. It is simply given by

$$L_C = L_E + L_D. \qquad (6.10)$$

In many applications it is important to keep coding delay low, which implies that both L_E and L_D should be kept as small as possible. In the case of a backward-adaptive source coder, this also helps to keep the adaptation delay low.

6.6 Forward and backward adaptivity compared

The most obvious difference between a forward- and backward-adaptive source-coding system is that unlike the forward-adaptive system, the backward-adaptive system requires no transmission of parameter bits, also called side information bits. This seems to decrease the bit rate. However, apart from this advantage, there are also a few drawbacks.

As the parameters in a backward-adaptive coder are estimated from past decoder outputs, they may be partly wrong. This can have two causes. The first is that the decoder outputs are distorted due to quantization, which influences the outcome of parameter estimation. The second cause is that because of adaptation delay the parameters are computed for a time interval which lies before the samples they are applied to. The errors in the parameters can result in an increased distortion. Particularly if signal characteristics change rapidly as a function time, e.g. if abrupt changes occur within Q' samples, this may pose serious problems. If characteristics vary slowly, e.g. if a change takes several times the Q' samples, the use of a backward-adaptive system can be advantageous. Note that the parameters used in a forward-adaptive system may be also erroneous, as they have to be quantized before use. With regard to the correctness of the parameters, it seems to depend on signal statistics and application as to whether forward- or backward-adaptive coding is preferable.

Uncorrected channel errors can result in different parameter sets in the source encoder and decoder. This is true for both backward- and forward-adaptive systems. However, in a backward-adaptive system an uncorrected channel error will always lead to a parameter mismatch, whereas in a forward-adaptive systems this only happens if parameter bits are affected. In a backward-adaptive system parameter mismatch may increase distortion for some time as new parameter sets are derived from past decoder outputs. This phenomenon is called error propagation in time. To cope with it, a backward-adaptive source coder may require stronger error protection, which increases the final channel bit rate.

Another possible drawback of backward-adaptive systems is that they contain a feedback loop and therefore may become unstable. Due to the non-linear nature of parameter estimators, this may be difficult to analyse.

6.7 Adaptive transform coding

Adaptivity in transform coders mostly refers to adaptivity in quantization and coding of transformed input samples, the transform is usually a fixed discrete Fourier transform or discrete cosine transform. Only rarely is it adapted to the input signal, examples are given in [59] and Chapter 12.

In adaptive transform coders, the adaptation period K is usually equal to the block length N. In most cases the parameters are estimated either from the N input samples before transformation or from the transformed samples. This means that parameter estimation is usually non-recursive and that forward adaptation is used. As a result

$$K = N = Q \tag{6.11}$$

Examples are discussed in Chapter 12. Practical values of block lengths range from 128 to 1024 for speech and music coding, in picture-coding systems blocks of 4×4 up to 16×16 are used.

It is interesting to see the effect of block length on adaptation and coding delay. Only the one-dimensional case is considered. It is assumed that parameter estimation is instantaneous, i.e. $L_P = 0$, which is usually the case. The encoding delay satisfies $L_E = N - 1$ and $L_D = 0$. Note that the latter relation is only true if transmission is instantaneous. According to (6.8), (6.11) and the above remarks on estimation and encoding delay, the adaptation delay satisfies

$$L_{AF} = 0, \tag{6.12}$$

and the coding delay

$$L_C = L_E = N - 1. \tag{6.13}$$

If a transform coder is used in a backward-adaptive source coder, the adaptation delay is given by

$$L_{AB} = N. \tag{6.14}$$

In speech and music coders the sampling frequency is between 8 and 48 kHz. For commonly used block lengths of between 128 and 1024 samples, this would result in adaptation delays between 16 and 21 ms and an unacceptably slow tracking of the signal characteristics, which can change within 1 to 5 ms.

6.8 Adaptive subband coding

In subband coding the splitting into subbands is rarely adaptive. Adaptivity in this section refers to adaptivity in quantization and coding. Usually adaptive source coding is performed on the subband signals. Examples are APCM coding, as is shown in Subsection 6.10.3 and Chapter 8, or predictive coding, e.g. [56].

If the decimation factors in all subbands are equal to the number of subbands N, then for the adaptation period K, expressed in the number of input samples, we have

$$K \geq N. \tag{6.15}$$

If the decimation factors are not all equal, which is the case for a non-uniform subband splitter, e.g. [60], then

$$K \geq N_{\max}, \tag{6.16}$$

where $N_{\max}$ is the largest decimation factor. In that case it could be that parameters in subbands with decimation factors less than $N_{\max}$ are adapted more often. For reasons of convenience only the case of equal decimation factors is considered further. If source coding in the subbands is block-based, with block length N_{sub}, then

$$K \geq NN_{\text{sub}}. \tag{6.17}$$

The number of subbands N ranges from 2, e.g. the G.722 speech coder, to 32 for a high-quality music coder [61].

The parameters are usually either estimated from $N \cdot N_{\text{sub}}$ input samples before transformation [61] or from the transformed samples [60]. This means that parameter estimation is non-recursive and that forward adaptation is used. As a result

$$K = Q = NN_{\text{sub}}. \tag{6.18}$$

Coding delay of a subband coder depends on the length of the subband filters and on the block length N_{sub}. Here it is assumed that the delay of the encoder filter bank and the decoder filter bank are both given by $L_F/2$. Then the encoding delay satisfies

$$L_E = L_F/2 + NN_{\text{sub}} - 1, \tag{6.19}$$

and the decoding delay

$$L_D = L_F/2. \tag{6.20}$$

According to (6.8), (6.18) and (6.19), the adaptation delay satisfies

$$L_{AF} = L_P - L_F/2 \tag{6.21}$$

and the coding delay

$$L_C = L_F + NN_{\text{sub}} - 1. \tag{6.22}$$

If parameters are estimated instantaneously from the subband signal block, as is the case in the APCM coder of Example 5 on page 19, then automatically $L_P = L_F/2$ and

$$L_{AF} = 0. \tag{6.23}$$

In the case of parameter estimation from the input signals this can usually be achieved by adding a delay of $L_F/2$ samples before parameter estimation.

Backward-adaptive source coding can be used in combination with subband coding. Generally, only the additional source coding in the subbands is backward-adaptive, e.g. backward-adaptive predictive coding, as discussed in Section 6.9. The subband filters are almost never part of the feedback loop, because they greatly increase the adaptation delay.

6.9 Adaptive linear predictive coding

In the cases of transform and subband coding, source encoders are rendered adaptive to changes in the signal characteristics by changing quantizers or coders. In general, the transform or the subband filtering are not adapted. In the case of predictive coding the situation is different. If a straightforward realization of $A'(z)$ of Figure 5.14 on page 123 is used, the prediction coefficients $a_1, \ldots, a_p$ have to be chosen such that the signal's power spectral density is approximated by

$$\sigma_\epsilon^2 \frac{1}{|1 - A'(\exp(j\theta)|^2}.$$

If another filter structure, such as a lattice filter, is used, other filter parameters have to be adapted. There are many ways of doing this. An excellent review is given in [57]. Here only the straightforward example of Figure 5.14 on page 123 is considered in more detail. Linear prediction is most often used in speech coding systems. In that case, the sampling frequency is usually 8 or 16 kHz, the order of prediction varies from 8 to 16 and the block length Q from which parameters are estimated is between 128 and 256. Methods of computing prediction coefficients are given in [4, 11, 57].

An interesting feature of predictive coders is that the encoding and decoding filters do not introduce a coding delay. This is easily verified by cascading the encoder and decoder filters of Figure 5.13. This implies that the adaptation delay depends only on the effective parameter estimation memory size Q', on the adaptation period K and the parameter estimation delay L_P. For practical methods of estimating prediction coefficients, the latter equals zero. In a forward-adaptive predictive coder, the adaptation delay can only be made zero by adding additional encoding delay. In that case, prediction coefficients are usually estimated non-recursively from a block of Q input samples. This implies that

$$K = Q = N, \tag{6.24}$$

and that, to make the adaptation delay zero, a coding delay of Q samples has to be introduced. In that case, the total coding delay is Q samples. If the quantization is block-based, as is the case for APCM or vector quantizer, then the coding delay may increase further. If the coding delay is required to be zero, then the adaptation delay will be $Q - 1$ sample periods. This can be reduced, if a lower value of K is chosen, but at the price of a higher bit rate because of the increased parameter bit rate.

In the case of backward adaptation, prediction coefficients are estimated from the decoded output signal $\hat{x}[n]$, and in a recursive backward-adaptive system also from the previous set of prediction coefficients. According to (6.9), assuming that $L_P = 0$, it follows that the adaptation delay depends only on Q' and K. Assume that N decoded output samples are used to derive prediction coefficients. Since in this case no parameter information is transmitted, it pays to decrease the adaptation period significantly, so that $K \ll Q$. As a result, the coding delay can be zero, and the adaptation delay is slightly more than $N/2$. This makes backward-adaptive predictive coding particularly suited for situations where the coding delay has to be very short. Note that because of the increased adaptation rate the complexity of source encoder and decoder has also increased.

6.10 Bit-rate control

6.10.1 Variable bit rates

The output bit stream of a source encoder does not always have a fixed rate. For instance, in Subsection 2.2.2 it was mentioned that a further reduction in bit rate can be obtained by assigning short codewords to quantizer output symbols with a high probability and long codewords to those with a low probability. In this way the average bit rate is reduced but the bit rate is no longer constant as a function of time. It can also happen that quantization is in some way dependent on the signal and does not operate with a fixed number of levels, which also results in a variable bit rate.

In transmission systems a fixed bit rate is often required or there may be restrictions, e.g. on the maximum bit rate. In the following it is assumed that the bit rate of the output of the source encoder is varying and that the channel only accepts a fixed bit rate. The varying bit rate is converted to a constant one. Two methods of converting a variable bit rate to a constant one are discussed in the following subsections: buffer control and adaptive bit allocation.

6.10.2 Buffer control

The bit rate can be controlled by buffers. A buffer is a memory device from which bits are read at a constant rate and to which bits are written at a variable rate or vice versa. A buffer with a variable input rate and a fixed output rate is placed after the coder block in the source encoder. If a buffer is used in the source encoder, the symbol rate at the output of the decoder block will not be constant. Consequently, the output sample rate of the source decoder is not constant. This is usually undesirable, therefore in the source decoder another buffer has to be used to regularize the output sample rate. This buffer can be placed at several points: before the decoder block, between decoder and reconstruction block, between reconstruction block and inverse transformation block, and after the inverse transformation block. Note that the buffer before the decoder block is one with a fixed input rate and a variable output rate. The buffer at any other position has a variable input rate and a fixed output rate.

If, per unit of time, more bits are written into a buffer than are read from it, the number of bits stored in the buffer will increase as a function of time. If this situation continues, all memory locations will be used. If the input bit rate continues to be greater than the output bit rate, buffer overflow will occur. The consequence is that buffer-output bits will be lost.

If per unit of time less bits are written into a buffer than are read from it, the number of bits stored in the buffer will decrease as a function of time, until the buffer is empty. The next output bit cannot be read. This is called buffer underflow.

In the following only the buffer in the source encoder is considered. Assume that it is filled half-way. Then, if it is large enough, there should be enough bits stored in the buffer to ensure a fixed buffer-output bit rate when the buffer-input rate drops temporarily. Also, there should be enough empty locations in the buffer to temporarily store input samples when there is a peak in the buffer-input bit rate. Unfortunately, in general no guarantee can be given that neither buffer underflow nor buffer overflow are excluded. To deal with this problem, buffer control is used.

In a buffer-control system the quantization block is made dependent on the number of memory locations used in the buffer. This number is called the buffer status. For instance, if a threatening buffer overflow is detected, the input rate is decreased by applying coarser quantization. If a threatening buffer underflow is detected, the input rate is increased by applying finer quantization. Since the coding block depends on the output statistics of the quantizer(s) it is sometimes also adapted.

There are many ways of making the quantization block dependent on the buffer status and no general rules for this can be given. Note that in the case of threatening overflow the distortion may increase due to coarser quantization and that in the case of threatening underflow the distortion may decrease because of finer quantization. In spite of the fact that buffer control is very important in practical source-coding systems, not much about it is found in source-coding literature.

In order to make the right reconstruction, the source decoder must know how the quantization block is adapted as a function of time. In principle there are two ways to achieve this. The first is that this information is transmitted as side information. The second is that the source decoder derives the buffer status from the incoming bit stream. Both ways are applied in practice, the first one is less sensitive to undetected channel errors, but requires a somewhat higher bit rate. Sometimes an intermediate form is used, where the buffer status is transmitted from time to time. Figure 6.11 shows a source encoder equipped with buffer control.

Buffer control is a non-linear feedback operation. If it is not carefully designed, it runs the risk of being unstable. Furthermore, its performance depends on the signal's statistics. It decreases probabilities of buffer overflow and underflow. However, if signal statistics do not match the assumptions on which quantization, coding and buffer control are based, buffer overflow and underflow may still occur. The buffer-control algorithm must contain emergency measures to prevent those situations.

146 Adaptive source coding

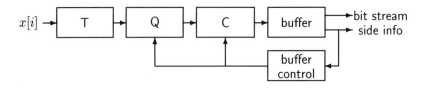

Figure 6.11 *Source encoder with buffer control. Side information is not always transmitted.*

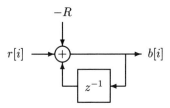

Figure 6.12 *Writing to and reading from the buffer memory.*

The emergency measure that prevents buffer overflow forces the buffer-input bit rate to be less than the buffer-output bit rate. This can be achieved by not transmitting part of the data, or by further reducing the quantization accuracy. The emergency measure that prevents buffer underflow generally consists of adding stuff bits to the transmitted data. The source decoder is informed about what has happened in the ways that have been described above.

In the remainder of this subsection, an example of a simple buffer control procedure will discussed in greater detail.

Example 27 We consider a simple two-dimensional transform coder, used in a picture-coding system. The source encoder consists of a transformation block, a quantizer which quantizes all transform coefficients uniformly with step size Δ and a Huffman coder.

The buffer contents when the ith block has been coded will be indicated as $b[i]$. Let us assume that the number of bits used to code this block is $r[i]$ and that the desired buffer-output bit rate is R bits per block of samples. The buffer content $b[i]$ can then be expressed as

$$b[i] = b[i-1] + r[i] - R, \tag{6.25}$$

This recursive relation is shown in Figure 6.12.

In the buffer-control procedure of this example, attempts are made to control the quantization step size in such a way that the buffer content is as constant as possible around a certain value B_{ref}. Usually this value is half of the maximum contents of

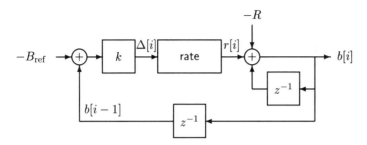

Figure 6.13 *Buffer control assuming a proportional relation between the quantizer step size and $(b[i-1] - B_{\mathrm{ref}})$.*

the buffer. A control parameter has to be chosen that expresses the deviation of the buffer content from the desired value. This control parameter is used to determine the quantization step size $\Delta[i]$ with which block i is quantized. An obvious control parameter is the difference $d[i]$ between the buffer contents after coding the $(i-1)$st block and B_{ref}, given by

$$d[i] = b[i-1] - B_{\mathrm{ref}}. \tag{6.26}$$

A procedure has to be defined to determine the step size $\Delta[i]$ from $d[i]$. A straightforward method is to assume a linear relation

$$\Delta[i] = kd[i] + m \tag{6.27}$$

between the step size $\Delta[i]$ and the control parameter $d[i]$. In this relation m is a constant, which, without loss of generality, will henceforth be assumed to be zero. This relation is schematically indicated in Figure 6.13. The box named 'rate' computes the bit rate as a function of the step size. In general, this function is non-linear. Figure 6.14 shows an example of such a function. The parameter k in Figure 6.13 will be called the control gain. Generally, the value of k will be determined from experiments with a large set of pictures.
◇

Some final remarks about buffer control have to be made. Firstly, the above example gives only a very first impression of the design of a buffer-control procedure. Secondly, in reality more complex buffer-control procedures are sometimes used, in which not only the difference between the wanted and the true buffer contents is used as a control parameter, but also the difference between the wanted and the true bit rate per block.

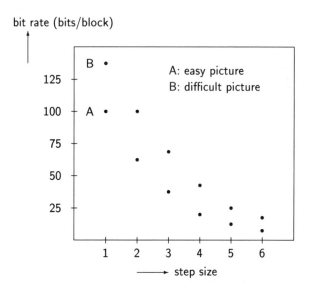

Figure 6.14 *Typical relation between step size and number of bits per block of samples.*

6.10.3 Adaptive bit allocation

Adaptive bit allocation can be regarded as an example of forward-adaptive quantization as well as a method of forward-adaptive bit-rate control, in which sense it can be an alternative to buffer control. It has the advantage over buffer control that instabilities cannot occur. However, transmission of side information is always required and complexity may be higher.

It is assumed that the transformation block splits up the signal into parts which are quantized separately. This happens, for instance, in a subband coder for speech or audio signals. Furthermore, it is assumed that the coder does not perform any further bit-rate reduction, but only maps the quantizer output symbols on to binary codewords. The bit rate R is then directly determined by the numbers of possible output symbols of the quantizers. The following example describes an adaptive bit-allocation procedure in a very simple subband coder. A more extensive example of adaptive bit allocation is described in Chapter 8.

Example 28 Figure 6.15 shows a simple subband coding scheme. Before quantization and coding, a signal is split into frequency bands called subbands. The boxes LP_E, HP_E, LP_D and HP_D are lowpass (LP) and highpass (HP) filters. The lowpass filters have a passband from zero to one quarter of the sampling frequency, the highpass filters have a passband from one quarter of the sampling frequency up to half the sampling frequency. After the filters in the source encoder, the signals are decimated by a factor of two. The

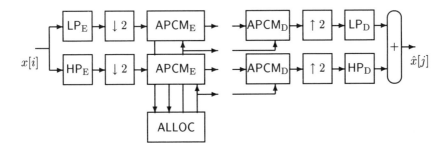

Figure 6.15 *Simple subband source encoder and decoder with adaptive bit allocation.*

signals after the decimators are the lowpass and highpass subband signals. Before the filters in the source decoder, the sampling rate is doubled by inserting zeros. If the filters are carefully designed, and if there is no further quantization, a perfect, or almost perfect, reconstruction of the input signal can be obtained at the output [8, 54, 55]. The blocks APCM$_E$ and APCM$_D$ are APCM encoders end decoders with uniform quantizers, as have been described in Example 5 on page 19. This is an example of a simple two-band subband coder, usually the number of subbands is higher [4, 60].

It is further assumed that the number of quantizer output symbols can be expressed as

$$M_{y,L} = 2^{b_L}, \quad M_{y,H} = 2^{b_H}.$$

The subscripts L and H refer to lowpass and highpass subbands, respectively. The APCM coder is blockwise adaptive, as is the case in Example 23 on page 132, but now the numbers b_L and b_H, and thus the quantization accuracy, are also allowed to vary per block of N samples. To enable correct decoding, they must be coded and transmitted as side information to the source decoder. This increases the bit rate. The total bit rate is given by

$$R = b_L + b_H + \frac{2\lceil \log_2(M_x) \rceil + 2\lceil \log_2(M_b) \rceil}{N},$$

where M_b is the number of bits used to code b_L and b_H.

A constant bit rate implies that

$$b_L + b_H = B \tag{6.28}$$

The block ALLOC has the task of choosing b_L and b_H in such a way that the bit rate is constant and the distortion minimal. Assume that the mean-square error is a good distortion measure. In that case, it is fairly easy to compute the b_L and b_H analytically. The total distortion is given by

$$D = \frac{\Delta_L^2}{12} + \frac{\Delta_H^2}{12}, \tag{6.29}$$

150 Adaptive source coding

where Δ_L and Δ_H are the quantization step sizes, given by

$$\Delta_L = \frac{2\hat{x}_{|\max|,L}}{2^{b_L}}, \quad \Delta_R = \frac{2\hat{x}_{|\max|,R}}{2^{b_R}}. \tag{6.30}$$

On substitution of (6.30) into (6.29) one obtains

$$D = \frac{\hat{x}^2_{|\max|,L} 2^{-2b_L}}{3} + \frac{\hat{x}^2_{|\max|,R} 2^{-2b_R}}{3}. \tag{6.31}$$

Lagrange optimization of D under the constraint (6.28) yields

$$b_L = \frac{B + \log_2(\hat{x}_{|\max|,L}) - \log_2(\hat{x}_{|\max|,H})}{2} \tag{6.32}$$

$$b_H = \frac{B + \log_2(\hat{x}_{|\max|,H}) - \log_2(\hat{x}_{|\max|,L})}{2}.$$

Of course the result is non-integer, but in this simple case of two subbands, rounding the result gives a solution which still satisfies (6.28). The block ALLOC can, for instance, compute the b_L and b_H by means of a lookup table. Note that the bit allocation is adaptive to the signal, since it depends on the values b_L and b_H. Note that problems may arise when one of the peak values is substantially larger than the other. These problems can be avoided by adding the constraints $b_L \geq 0$ and $b_H \geq 0$, but this would make the optimization problem more complicated.
◇

In the above example it can be seen that the signal is split into components which are quantized separately. Each quantizer output maybe coded with a different bit rate. The total bit rate is the sum of the bit rates of all quantizer outputs plus the bit rate needed for side information. Distortion is controlled by choosing the number of quantization levels for each quantizer adaptively to the signal. This is the general idea behind adaptive bit allocation. Note that a certain amount of storage is required to analyse the signal before the allocated bits can be computed, but only in the source encoder. Figure 6.16 shows a more general basic diagram of a source encoder with adaptive bit allocation.

6.11 Problems

Problem 30 Rewrite the parameter estimation method of Example 25 in a recursive form.

Problem 31 Derive a relation between q in Example 25 and α in Example 24 for the same Q'.

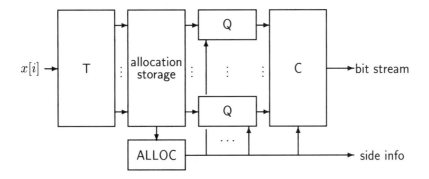

Figure 6.16 *Source encoder with adaptive bit allocation.*

Problem 32 Assume that in Example 28 the distortion in the lower subband is considered more disturbing than in the higher subband. In that case one can express this in the total distortion by introducing weighting coefficients. In that case, (6.29) is replaced by

$$D = w_L^2 \frac{\Delta_L^2}{12} + w_H^2 \frac{\Delta_H^2}{12}, \tag{6.33}$$

where w_L and w_H are positive weighting factors with $w_L > w_H$. Compute b_L and b_H for this case. Hint: Use the previous result. Avoid doing the Lagrange optimization yourself.

Part II

Applications

7
Introduction to Part II

Part I of this book has dealt with the theoretical aspects of source-coding systems for digital signals. Chapters 1 and 2 have explained that a source-coding system can be thought of as being composed of a number of functional blocks, called transformation, quantization and coding blocks. Each of these blocks has been discussed in detail in Chapters 3–5. The adaptation of these blocks to the characteristics of the input signal has been dealt with in Chapter 6.

Part II of this book illustrates how the principles explained in Part I can be applied so as to reduce the bit rate of real-life digital signals. It is roughly divided into two parts. One part is concerned with source coding for digital audio signals and the other with source coding for digital pictures. In contrast to Part I, Part II does not use examples to explain or introduce theoretical concepts. In fact, Part II can be viewed as consisting only of examples of source-coding systems. Moreover, the explanations given in Part II are not very theoretical.

Chapter 8 of Part II is about subband coding of digital audio signals. Using the method explained in that chapter, the bit rate of compact disc audio signals can be reduced by more than a factor of four without introducing audible errors. A similar principle is applied in the recently introduced Digital Compact Cassette (DCC®) system.

Chapters 9–14 describe a large variety of methods for the source coding of digital pictures and Chapter 15 describes a method for the source coding of picture sequences. Examples are given of systems employing pulse-code modulation (PCM), differential pulse-code modulation (DPCM), transform coding (TC), subband coding (SBC) and pyramidal coding (PC). Numerous photographs are provided to illustrate the performance of these systems.

It has been attempted to describe the applications in Part II as fully as possible with reference to the theoretical framework of Part I. Most of the chapters of Part II contain sections on transformation, quantization and coding, and references to Part I are given.

Readers with some experience in source coding of digital signals can read Part II independently. However, we recommend readers without the required experience to study Part I thoroughly first.

8
Subband coding of audio signals

8.1 Introduction

The transmission and storage of high-quality digital audio signals are becoming important for the audio industry. Examples are digital broadcasting, new applications for optical disks and tape recording of digital audio signals. The bit rate of a high-quality stereophonic digital audio signal, as recorded on a compact disc is about 1.4 Mbits per second. For some transmission channels or storage media this is too high and therefore source coding is required. Since digital audio is associated with high quality, a perceptible loss of quality cannot be tolerated.

This chapter discusses some general aspects of source coding for digital audio signals. The source-coding method used in the Digital Compact Cassette (DCC®), called Precision Adaptive Subband Coding (PASC®), [62] is discussed in some more detail.

Source coding for digital audio signals will be explained for monophonic signals only. A simple extension to stereophonic signals consists in treating the left and right channels as independent monophonic signals, but there are other, more elaborate possibilities [51, 63]. The audio signal that is considered here is the compact disc signal. The sampling frequency is 44.1 kHz and each sample consists of 16 bits. There are, however, other commonly used standards in which other sampling frequencies and different numbers of bits per sample are used. PASC is therefore capable of coding 16-bit or 18-bit audio signals sampled at 32, 44.1 or 48 kHz.

Source coding of audio signals at low bit rates generally introduces errors. The source-coding method described in this chapter attempts to keep coding errors inaudible by exploiting simultaneous masking. This is the perceptual phenomenon that a weak signal, e.g. quantization noise, is masked (i.e. made inaudible) by a stronger signal, e.g. a pure tone in the audio signal. For reasons of convenience, the masking signal will henceforth be called the masker and the signal to be masked will be called the target. In order to be masked, the target's level must be below the so-called masking threshold. Simultaneous masking, also simply called masking, is briefly explained in Section 8.2.

Masking is most effective if both the masker and the target are within a rather narrow frequency band. This suggests the use of subband coding, in which the signal is split into frequency bands which are quantized and coded. Subband coding has been explained in Chapter 5. The structure of the PASC subband-coding system is given in Section 8.3.

In order to keep the quantization errors inaudible, quantization has to be such that the quantization noise is masked by the audio signal. This is achieved by using APCM source coding in the subbands. The subband signals are split into blocks. Each block is scaled to a unit level and then quantized by a uniform quantizer. Quantized data and peak values are transmitted. In this manner the power of the quantization noise can be controlled by allocating a certain number of bits to each quantizer. APCM is described in detail in Chapters 2 and 6.

In an audio subband-coding system we can distinguish in-band masking, which occurs when both masker and target are in the same subband, and out-of-band masking, which occurs when masker and target are in different subbands. Both are exploited in the method described in this chapter. Section 8.4 explains how the maximum quantization-noise power that is masked, which is further called the masked power, can be estimated for each subband.

Once the masked powers have been computed for all subbands, bits are allocated to the quantizers. Ideally, the number of bits for each quantizer should be such that in each band the quantization noise is completely masked. However, the masked powers are signal-dependent and therefore the number of bits needed to ensure complete masking varies with time. Coding systems with a fixed bit rate are considered here. The available bits must therefore be spread over the subbands in such a way that the audible degradation of the output signal is minimal. This requires an adaptive bit allocation method, which is described in Section 8.5.

Results obtained with PASC are discussed in Section 8.6. Section 8.7 briefly discusses other coding systems.

8.2 Masking

Masking means that a weak signal is made inaudible by a simultaneously occurring stronger signal. Masking has been discussed in great detail in [64, 65, 66]. The use of masking in subband coding has been described in, for instance, [60, 67, 68]. Results from [64] are repeated in the next two paragraphs to illustrate the masking effect.

Consider as a target one pure tone. This tone is inaudible if its sound-pressure level (SPL) [69, 70] is below the threshold of hearing. The threshold of hearing is shown as a function of frequency in Figure 8.1. In the presence of a second, stronger signal, the sound-pressure level above which the target is audible differs from the threshold of hearing. It is raised for frequencies close to the frequency of the stronger signal. This new threshold is called the masking threshold and the

158   Subband coding of audio signals

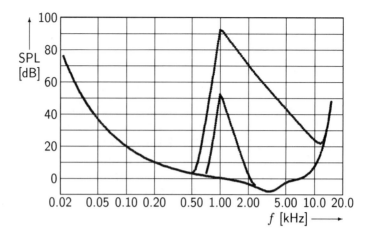

Figure 8.1   *Threshold in quiet (lower curve) and masking thresholds of narrow-band noise maskers centred on 1.0 kHz at sound-pressure levels $L_N = 100$ dB (upper curve) and $L_N = 60$ dB (middle curve) for tonal targets. After [65].*

second, stronger signal is called the masker. The target is masked if its sound-pressure level is below the masking threshold. This is illustrated in Figure 8.1, where masking thresholds of narrow-band noise signals with a bandwidth of 90 Hz, centred on 1 kHz, at various sound pressure levels L_N are shown.

The masking threshold as a function of frequency depends on the sound-pressure level and the power spectral density of the masker. Apart from a level difference and some irregularities at frequency f_m and its harmonics [64, 65], tonal maskers have masking thresholds that are similar to those for noise maskers. We assume here that they have identical shapes, but that the masking curve of a tonal masker is some 15 dB below that of a noise masker. Figure 8.2 shows masking thresholds of narrow-band noise maskers with different centre frequencies f_m. Masking thresholds of narrow-band noise maskers have different shapes for centre frequencies below and above 500 Hz, as is illustrated by Figure 8.2. The masking thresholds for both pure tone and narrow-band noise maskers show an asymmetry about the centre frequency of the masker. Signals with frequencies higher than the centre frequency of the masker are better masked than signals with frequencies lower than the centre frequency of the masker.

So far, masking of a pure tone by another pure tone or by a narrow-band noise signal has been explained to illustrate the masking effect. In a waveform coder the targets will be mainly noise targets, owing to quantization. Very little relevant information on masking of noise targets seems present in the literature, e.g. [71]. To analyse masking of noise targets, noise targets first have to be properly defined [72]. As the ear seems to integrate over limited frequency regions called critical bands [70], it seems reasonable to define critical bands of noise as targets. The definition

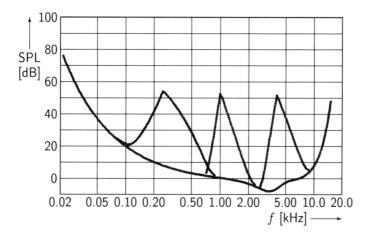

Figure 8.2 *Threshold in quiet (lower curve) and masking thresholds of narrow-band noise signals centred on frequencies $f_m = 250$ Hz (left-hand curve), $f_m = 1$ kHz (middle curve) and $f_m = 4$ kHz (right-hand curve) for tonal targets. The masker level is 60 dB in all cases. After [65].*

of critical bands given in [70] is used here. According to this definition, critical bands below 500 Hz are 100 Hz wide while above 500 Hz the critical bandwidth is approximately a third of an octave. This means that the ratio of the upper-edge cutoff frequency and the lower-edge cutoff frequency is about $2^{1/3}$. Consecutive critical bands are numbered from 1 to N. For audio signals sampled at a frequency $f_s = 44.1$ kHz, 26 critical bands have to be taken into account. The critical band scale is also used as a measure of frequency, called critical-band rate. The symbol used to denote it is z. The corresponding unit is the bark.

A critical-band noise target has a flat power spectral density within one critical band and a zero power spectral density outside outside that band. In fact, because of the integration one would expect the masking thresholds for critical-band noise targets to be similar to those for tonal targets. Figure 8.3 shows a masking threshold of a tonal masker of 400 Hz for critical-band noise targets as a function of the critical-band number. It is the result of an informal experiment, with one of the authors of this book as a subject. It confirms the expectation that masking thresholds for critical-band noise targets are similar to those for tonal targets. Although further study of masking thresholds for critical-band noise targets and tonal and noise maskers is needed, it is further assumed that the thresholds for tonal targets can be used instead. A second conclusion from the fact that the masking curves for tonal and critical-band noise targets do not differ very much, is that the precise spectral shape of the noise target within the critical band is not important. This is convenient because the noise targets in a subband-coding system are due to quantization noise, which, as is known, cannot always be considered spectrally flat.

160 Subband coding of audio signals

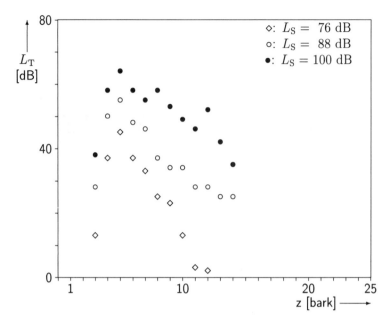

Figure 8.3 *Masking thresholds L_T of critical-band noise targets with a 400 Hz masker at various sound-pressure levels L_S.*

On the basis of the assumptions in the preceding paragraphs, Section 8.4 presents a simple, analytical masking model that can be used in an audio source-coding system.

8.3 Subband coding

It is clear from Figure 8.1 that masking is strongest for frequencies close to the frequency of the masking signal. This suggests that the masking phenomenon can be usefully employed in a subband-coding system. As has been explained in Chapter 5, the source encoder in such a system splits the signal into subbands, which are quantized and coded. The coded quantizer-output symbols are then transmitted, in this case together with some additional data. The source decoder reconstructs the subbands from the decoded quantizer-output symbols and additional data. The subbands are subsequently merged into a replica of the original signal.

The splitting of the signal into subbands and the merging of the subbands into a replica of the original signal are done by decimating and interpolating filter banks, such as quadrature-mirror or conjugate quadrature-mirror filter banks [55,

73]. Owing to the decimation, the sample rate of a subband signal is twice the subband's bandwidth, so that the total sample rate after splitting is the same as the sampling frequency at the input. The ratio of the sampling frequency of the input signal of the filter bank and the sample rate of a signal in a subband is called the decimation factor of that subband.

As can be seen from Figure 8.2, the 'bandwidth' of the masking threshold increases with the frequency of the masking signal. It would therefore be convenient if the bandwidths of the subbands were also to increase with frequency. This is the case in some proposals for subband coding, e.g. [60, 68]. In the PASC system, however, all 32 subbands have the same bandwidth, which equals the sampling frequency divided by 64.

In the process of quantization, quantization noise is added to the signals. If the filter banks have good frequency-separating properties, the quantization noise remains in the subband it was added to. It is masked by the audio signal if the signal-to-noise ratio is above a certain threshold. This means that the quantizers have to operate at a predetermined signal-to-noise ratio. As has been explained in Chapter 2, this can be achieved with APCM source coding. In an APCM source encoder the signal is first divided into blocks. The maximum absolute values of these blocks, called peak values, are computed. By dividing the samples in the blocks by the peak values, they are scaled to a unit level. The scaled blocks are then quantized with a uniform quantizer. The signal-to-noise ratio expressed in dBs is proportional to the number of bits that is used in the quantizer, so that the signal-to-noise ratio of a quantizer can be predetermined by allocating a certain number of bits to it.

Figure 5.10 on page 114 shows the general basic diagram of an N-band subband coder. In the PASC system $N = 32$ and quantizers and reconstructors are replaced by APCM encoders and decoders. This is illustrated in Figure 8.4. As can be seen, coded quantizer-output symbols as well as coded peak values and side information indicating the number of quantization bits are transmitted.

On the one hand, the number M of samples in the blocks should be chosen to be sufficiently small, so that changing statistics in the audio signal can be tracked and it can be assumed that the signal-to-noise ratio of the quantizer is constant. On the other hand, the number of samples in a block must not be too small because the bit rate then increases owing to the increased amount of side information. A maximum block length corresponding to 10 ms is found to give good results. In the PASC system a block length of 12 samples has been chosen. At an input sampling frequency of 44.1 kHz and a decimation factor of 32 this corresponds to a block duration of 8.7 ms.

In Section 8.4 it is shown how, for a given division of the signal into subbands, the masking effect described in Section 8.2 can be used to determine the masked powers in the subbands. It is also explained how the final bit rate depends on the division into subbands. Section 8.5 also shows how the number of bits allocated to each quantizer is computed.

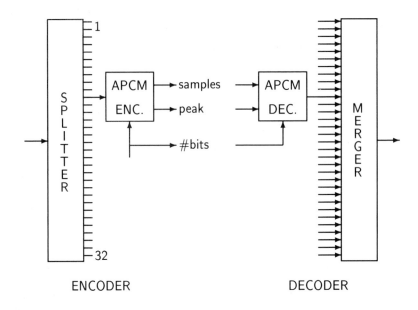

Figure 8.4 *Subband-coding system with 32 subbands.*

8.4 A masking model for subband coding

In Section 8.2 results reported in psycho-acoustical literature have been presented to illustrate the masking effect. These results cannot be used directly to compute masked powers in subbands, as is required in subband-coding systems. For instance, the masking of one pure tone by one other pure tone or by a narrow-band noise signal has already been discussed in Section 8.2. Music is a time-varying complex of tones combined with noisy signals. Although some experiments have been performed in that direction [74, 75], it is not known how the masking threshold of a complex of tones must be computed. In addition, the results of Section 8.2 are valid for only one tonal target. Quantization noises in subbands are essentially multiple noise targets. It is shown in [72] that a combination of individually masked targets may not always be masked. It is therefore unclear how, given a short-time music power spectral density, masked powers in subbands have to be computed.

As has been done in other audio source-coding systems [53, 68], a masking model is assumed which provides answers to the above problems. It must be emphasized that this model is partly based on assumptions that have never been completely justified. The masking model chosen here is simple. More elaborate models will lead to better coding systems but with increased complexity. As will be demonstrated in Section 8.6, good enough results can in fact be obtained with the simple model used here.

A masking model for subband coding

The masking threshold of a tonal masker is used as a basis for the masking model. It is assumed in this model that, on a bark scale, its shape is independent of the bark rate of the tone and of its sound-pressure level. In the model used here the masking threshold $t(z_m, z_t)$ of a pure tone masker at bark rate z_m with power σ_m^2 is approximated by

$$t(z_m, z_t) = \begin{cases} \sigma_m^2 \rho_{\max}(z_m) |z_m - z_t|^{-3}, & z_t \leq f_m, \\ \sigma_m^2 \rho_{\max}(z_m) |z_m - z_t|^{-10}, & z_t > f_m. \end{cases} \quad (8.1)$$

In this expression z_t is the bark rate of the target and $\rho_{\max}(z_m)$ the relative masking threshold at this bark rate, which depends slightly on the bark rate of the masking signal [67]. Figure 8.5 shows a stylistic approximation of the masking threshold according to the masking model of this chapter for a pure tone of 1000 Hz at a sound-pressure level of 90 dB. For a frequency of 1000 Hz, $\rho_{\max}(z_m)$ equals -20 dB [64, 65].

It is assumed that masking is additive: the masking threshold for a signal containing more than one frequency component can be obtained by adding the masking thresholds of the components. That this is permissible can be concluded from results in [74], though some of these have not been confirmed in [75]. From an engineering point of view the assumption of additive masking is very attractive. Otherwise, a straightforward computation of masked powers would not be possible.

In [72] it was found that a combination of individually masked targets may not always be masked, but that this combination of targets is again masked if a safety margin below the masking threshold of 9 dB is adhered to. This can be achieved by assuming that $\rho_{\max}(z_m)$ equals 30 dB rather than 20 dB.

As has been argued, the masking model used here is a simplification of reality. Coding systems based on it may show unexpected and unwanted effects. To avoid this, they have to be tested and optimized in extensive listening experiments.

After this discussion of the masking model used, it will now be shown how masked powers in subbands are computed. In computing the masked power in the subbands, three kinds of masking are considered: in-band masking, which is masking within a subband, and the masking of targets in subbands at higher and at lower frequencies. For all cases the masked power is computed as a function of the powers of the subband signals.

First it is assumed that there is only one signal in the subband with index i. This subband ranges from bark rates $z_{l,i}$ to $z_{u,i}$. The quantization noise is assumed to have a flat power spectral density in the subband. The worst-case situation for in-band masking occurs when the masker is a pure tone with a bark rate $z_{u,i}$. The minimum masking threshold in subband i due to the masker in subband i is given by $t(z_{u,i}, z_{l,i})$ defined in (8.1). This situation is illustrated in Figure 8.6.

The worst-case situation for the masking of noise in subbands at higher frequencies occurs when the masker in subband i is a pure tone with a bark rate $z_{l,i}$. The minimum masking threshold in subband i due to the masker in subband i is given by $t(z_{l,i}, z_{u,j})$ defined in (8.1). This situation is illustrated in Figure 8.7.

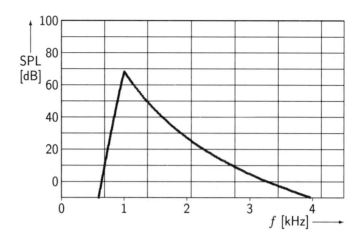

Figure 8.5 *Masking threshold as a function of frequency for a 1 kHz tonal masker with a level of 90 dB. Subband boundaries are indicated in the diagram.*

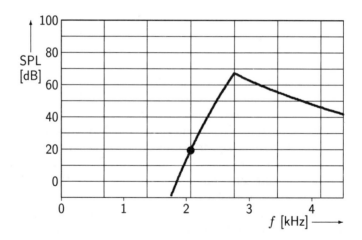

Figure 8.6 *Computation of in-band masking level for subband 4. Subband boundaries are indicated in the diagram. The level of the masker is 90 dB. The in-band masking level in subband 4 is indicated by a •.*

Derivation of the masking threshold for a target in a subband below the subband containing the masker is left to the reader. The contribution of each subband to the masked powers in all subbands can be computed in the manner described above. Because masking is assumed to be additive, the masked power in a subband can be obtained by adding the contributions of all subbands to the hearing threshold in that subband.

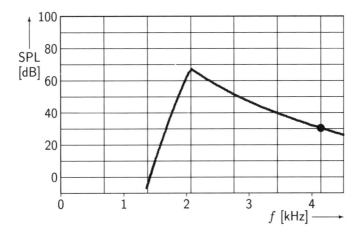

Figure 8.7 *Computation of masking level in subbands at higher frequencies. Subband boundaries are indicated in the diagram. The contribution of subband 4 to the masking level in subband 6 is indicated by a •. The level of the masker is 90 dB.*

The lowest achievable bit rate at a particular quality level depends on the division into subbands. The computations of the masked powers are based on worst-case assumptions. The real masked powers can be substantially higher. If the subbands are narrower, the results of these computations will, on average, be closer to the real masked powers. This effect, however, is limited for in-band masking, because in reality the tops of the curves of Figures 8.5, 8.6, and 8.7 are flatter than shown [67], so that there is no point in decreasing the bandwidth of the subbands below a certain value. Because of the effect mentioned here, narrower subbands lead to higher masked powers and consequently the number of bits required for quantization can be lower.

The results of this section are valid only if the distribution of signal power over the subbands is constant as a function of time. In reality that is not the case. In a subband-coding system the masked powers must therefore be computed periodically. As a consequence of the varying powers in the subbands, the number of bits needed to quantize each subband in such a way that the quantization noise is masked will also vary with time. The allocation of bits to the quantizers in such a way that the coding system produces a fixed bit rate is discussed in Section 8.5.

8.5 Adaptive bit allocation

Before the subband signals are quantized they are divided into blocks. As previously stated, the block size in the PASC system is 12 samples. The blocks are arranged in an allocation window which encloses all subband samples contained in a period of time. In the case of PASC operating at a sample frequency of 44.1 kHz this period is 8.7 ms.

In a subband coder with evenly spaced subbands, the number of samples N_w in an allocation window is given by the product of the number of subbands and the block length. In the case of PASC this figure is $32 \times 12 = 384$. If the desired bit rate at the output is R bits per sample, then a number of $N_w R$ bits must be distributed over the allocation window. The bit rate of PASC is 192 kbits/s. At a sampling frequency of 44.1 kHz, this means that $R = 4.354$ bits per sample. The number of bits available for one allocation window is given by $N_w = 1671$. A certain number of bits are used for peak values and side information describing the bit allocation. Good results have been obtained with peak values which are logarithmically quantized and coded with 6 bits and 4 bits of side information per block indicating the number of bits per sample used in the block. All subbands require bit-allocation information, so that 1543 bits are left for peak values and samples. Quantization is always performed with a uniform midtread quantizer which has an odd number of levels. This means that the possible number of quantization levels for a subband block is given by

$$1, 3, 7, \ldots, 2^{16} - 1.$$

Assigning one level to a subband block means setting all values in the block to zero. Only the bit allocation has then to be transmitted. Bit allocation is coded as a 4-bit number b, ranging from 0 to 15. In this case only the value $b = 0$ is transmitted. In all other cases the number of levels is given by $2^{b+1} - 1$. Each sample quantized with $2^{b+1} - 1$ levels is coded with $b + 1$ bits. If a certain subband block is quantized with more than one level, bits have to be transmitted for quantized and coded samples as well as for quantized and coded peak values. In the case of PASC coding a subband block quantized with $2^{b+1} - 1$ levels requires transmission of $12 \times (b + 1) + 6$ bits.

In the paragraphs which follow, a procedure is described for distributing the remaining bits optimally over the subbands, given a certain fixed bit rate or, in other words, given a certain number of bits for quantizing and coding peak values and samples in an allocation window.

First, the peak value and the power are computed for each block in the allocation window. The power in a block is obtained as the average of the squares of the samples. Using the masking model of Section 8.4, the masked powers are now computed for every block in the allocation window.

Assume that the blocks in the allocation window are numbered from 1 to 32. The estimated masked power in the ith block is denoted by $\sigma_{m,i}^2$ and the peak value by p_i. If the samples in the block are uniformly quantized and coded with b_i bits, then the power of the quantization noise in the ith block is assumed to be approximated by [4]

$$\frac{1}{12} \left(\frac{2 p_i}{2^{b_i}} \right)^2.$$

The noise-to-mask ratio in the same block is defined by

$$\frac{1}{12}\left(\frac{2p_i}{2^{b_i}}\right)^2 \frac{1}{\sigma_{m,i}^2}.$$

In what follows it will sometimes be more convenient to use a logarithmic noise-to-mask ratio γ_i defined by

$$\gamma_i = \log_2\left(\frac{p_i}{\sigma_{m,i}}\right) - b_i.$$

The adaptive bit allocation procedure is such that the total noise-to-mask ratio, given by

$$\sum_{i=1}^{32} \frac{1}{12}\left(\frac{2p_i}{2^{b_i}}\right)^2 \frac{1}{\sigma_{m,i}^2}$$

is minimized subject to the constraint that the total number of bits needed to code quantized samples and peak values does not exceed a certain number. In the case of PASC this number is 1543. A further constraint is that all b_i must be integers with $2 \le b_i \le 16$.

An elegant solution to this constrained integer-minimization problem can be derived from the theory given in [76]. The result of this procedure for the present case, as described below, is intuitively attractive. Initially, it is assumed that all blocks are quantized with one level. The number of bits then left to be divided over the allocation window is then 1543 and the logarithmic noise-to-mask ratio for the ith block is given by

$$\gamma_i = \log_2\left(\frac{p_i}{\sigma_{m,i}}\right).$$

Then the following four steps are repeated until there are not enough bits left to assign to a subband:

1. The block i with the highest logarithmic noise-to-mask ratio γ_i is searched for.
2. If no bits have been previously assigned to this block, the number b_i of bits per sample for this block is increased by two and the number of remaining bits is decreased by $2 \times 12 + 6 = 30$.
3. Otherwise, the number of bits per sample b_i for this block is increased by one and the number of remaining bits is decreased by 12.

If this procedure is started with sufficiently many bits, the result is that all noise-to-mask ratios will be approximately equal. The procedure can be refined by setting lower bounds to the noise-to-mask ratios of the subband. The reason for doing so is that, once the noise-to-mask ratio in a block is below a certain threshold, it is no longer necessary to add bits to the block because the quantization errors have become inaudible. Bits can thus be saved for other blocks.

Note that after the allocation procedure a small number of bits usually remain. The actual bit rate will therefore be a fraction less than the desired bit rate.

8.6 Results

The PASC system has been designed for coding stereophonic 16- or 18-bit digital audio signals with sampling frequencies of 32, 44.1 or 48 kHz. It has been extensively tested at a great number of test sites. Subjective blind tests with a large number of subjects have been performed. The outcome is that, at the relatively high bit rate of 192 kbit/s for each channel, the decoded signal can be considered indistinguishable from the original. This is almost always the case, even at lower bit rates of about 128 kbits/s per channel. If signal-to-noise ratios are measured, they are found to be between 15 and 30 dB, showing that the signal is heavily distorted.

8.7 Other systems

PASC is essentially a special variant of a 32-band subband audio-coding proposal which has been accepted as an ISO standard [63, 77]. In its most complex forms this system is capable of coding signals with a high quality at 96 kbits/s per channel and with an intermediate quality at 64 kbits/s per channel. The more complex versions have larger allocation windows containing 1192 samples. With more elaborate coding, and sometimes additional transformations, they further reduce the bit rate for samples, bit-allocation information and peak values. The masked powers are not derived from the powers in the subbands but by means of a discrete Fourier transform applied to the input samples. The masking models used are also somewhat more extensive.

A different kind of audio source coding is transform coding, e.g. [78, 79]. The difference is, of course, the transformation, which is a discrete cosine or a discrete Fourier transform. The transform block-length varies from 256 to 1024 samples. To avoid blocking effects, overlapping transforms [49], which can be regarded as special versions of subband filters, are used. The block length sometimes is made adaptive to the signal [59, 63, 80]. In that case transients are coded with shorter blocks. Most transform-coding systems for audio employ masking models similar to those used in subband coding. Differences can be found in the way transform coefficients are quantized, compared with how subband samples are quantized. Unlike in subband-coding systems, some type of variable-length coding, e.g. Huffman coding is generally used in transform-coding systems.

9

Picture source coding

9.1 Introduction

Chapters 10–15 are about source coding of digital pictures. This chapter gives an introduction to those chapters. Section 9.2 will describe what digital pictures are and how they are obtained, Section 9.3 will explain why source coding of pictures is needed, and Section 9.4 will give a brief summary of picture source-coding techniques which will be described in Chapters 10-15.

9.2 Digital pictures

Objects in the world surrounding us produce or reflect radiation, such as light or x-rays. A picture is a projection of the intensity of this radiation on a flat, usually rectangular, area. The world is characterized by three spatial coordinates and one time coordinate. The picture has only two spatial coordinates, which are generally referred to as the horizontal and vertical coordinates. How is this projection effected?

Let us consider the case that we want to record light, and let us assume, first, that only the luminance of the light has to be recorded. Luminance is a physical quantity describing what our human visual system perceives as brightness. Normally, a lens is used to focus on a flat, rectangular, area the light that is radiated or reflected by objects. The luminance image on this area can be considered to be a real-value function $L(x,y,t)$ of spatial coordinates, x and y, and time coordinate, t. The function $L(x,y,t)$ is non-negative, and it can be assumed that it is bounded, i.e. $0 \leq L(x,y,t) \leq L_{\max}$. Furthermore, it is usually assumed that $L(x,y,t) = 0$ outside the rectangular area.

A digital picture is obtained by sampling $L(x,y,t)$ in space, at a certain time instant $t = t_0$, and quantizing each of the samples. A rectangular sampling raster such as shown in Figure 9.1 is generally used. The resulting digital picture can be seen as a matrix of samples. These samples are sometimes called picture elements,

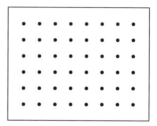

Figure 9.1 *Rectangular sampling raster.*

or simply pixels, or pels. For example, a CCIR 601 TV picture consists of 576 rows, which are usually called lines, with 720 luminance samples per line. Each sample is uniformly quantized with 256 quantization levels and 8 bits are used to represent the quantized samples. This means that the luminance is represented by integer numbers between 0 and 255. A value of zero corresponds to black, and 255 to white. Figure 9.2 shows the luminance of a digital CCIR 601 TV picture. The original photograph belongs to the CCETT France, and was recorded by Mr Geffroy of CCETT. The picture of Figure 9.2 will be used as the test picture throughout Chapters 10–15.

The recording of a colour picture is very similar to that of a monochrome picture, except that the incoming light is first divided into three colour components, which are then separately sampled. Commonly used colour components are red (R), green (G) and blue (B). To achieve compatibility with monochrome television signals, the colour components are usually first converted to a luminance component, often indicated as Y, and two so-called colour difference or chrominance signals. In the following chapters only monochrome digital pictures will be considered, and the luminance of the nth sample on the mth line of such pictures will be indicated as $p[m,n]$.

In many applications not a single picture but a sequence of pictures is recorded. Such a sequence is usually created by sampling the luminance and the colour components at regularly-spaced time instants. For example, the CCIR 601 TV signal consists of 25 pictures per second for Europe, and 30 pictures per second for the USA and Japan. The bit rate of this signal equals 166 Mbit/s. Each individual picture is sometimes called a frame, and the number of pictures per second is called the picture rate or frame rate. A sequence of pictures will be indicated by $p_i[m,n]$, where i indicates the index of a picture in the sequence.

In a sequence of digital pictures the sample values $p_i[m,n]$ can vary as a function of the discrete spatial coordinates m and n, and the discrete time coordinate i. The spatial and temporal spectral behaviour of $p_i[m,n]$ is often characterized by means of the Fourier transform.

Figure 9.2 *Example of luminance of CCIR 601 TV frame.*

For a one-dimensional signal $x[n]$ the Fourier transform is defined as

$$X(e^{j\theta}) = \sum_{n=-\infty}^{\infty} x[n]e^{-j\theta n}, \qquad (9.1)$$

where θ is the normalized frequency in radians. $X(e^{j\theta})$ describes the presence of frequency components with frequency θ in the signal $x[n]$. Since $X(e^{j\theta})$ is periodical with period 2π, it suffices to consider it for θ in the interval $[-\pi, \pi)$, the so-called fundamental interval.

Similarly, the three-dimensional Fourier transform of digital picture sequences can be defined as

$$P(e^{j\theta_h}, e^{j\theta_v}, e^{j\theta_t}) = \sum_{m=-\infty}^{\infty} \sum_{n=-\infty}^{\infty} \sum_{i=-\infty}^{\infty} p_i[m,n]e^{-j(\theta_h n + \theta_v m + \theta_t i)}. \qquad (9.2)$$

In this expression θ_h represents the horizontal spatial frequencies in the picture sequence, θ_v the vertical spatial frequencies and θ_t the temporal frequencies. $P(e^{j\theta_h}, e^{j\theta_v}, e^{j\theta_t})$ indicates to what extent frequency components with frequencies θ_h, θ_v, and θ_t are present in the picture $p_i[m,n]$.

Spatially, pictures always have a limited size. Assume that $p_i[m,n]$ only has non-zero values for $m = 0, \ldots, M-1$, and $n = 0, \ldots, N-1$, where M is the number of lines in a picture, and N the number of samples per line. Expression (9.2) then reduces to

$$P(e^{j\theta_h}, e^{j\theta_v}, e^{j\theta_t}) = \sum_{m=0}^{M-1} \sum_{n=0}^{N-1} \sum_{i=-\infty}^{\infty} p_i[m,n]e^{-j(\theta_h n + \theta_v m + \theta_t i)}. \qquad (9.3)$$

A large value of $P(e^{j\theta_h}, e^{j\theta_v}, e^{j\theta_t})$ at the higher spatial frequencies θ_h or θ_v corresponds to rapid changes of the luminance in a picture in the horizontal and the vertical directions, respectively. A typical example is the luminance change occurring at so-called edges, i.e. sharp luminance transitions at the border of different objects. A large value at the higher temporal frequencies θ_t corresponds to rapid luminance changes with time.

9.3 The need for picture source coding

During the last few decades, the number of applications in which pictorial data has to be stored or transmitted has rapidly increased. For example, the number of television (TV) signals transmitted to homes has increased from one or two to several tens. Another example is the videophone, which will be introduced in the near future. This device enables audiovisual communication via telephone channels. In the past, pictorial data such as television signals were recorded and transmitted in an analog form. At present, a shift towards digital representation can be observed.

Digital representation of pictures has a number of important advantages. Firstly, it enables a significantly higher picture quality to be obtained. Secondly, digital pictures can be more easily processed than analog ones. Over the last few decades a variety of digital signal processing techniques has been developed. As a result, digital pictures can be handled much more flexibly than analog pictures.

Representing pictures digitally also has a disadvantage, namely that their bit rate is high. The capacity required for their storage and the bandwidth required for their transmission is therefore also high. For example, standard-definition television signals (SDTV) of the so-called CCIR 601 format[81] have a bit rate of 166 Mbit/s. In the near future, television signals with an even higher resolution than TV signals will be introduced. These signals are referred to as high-definition television signals (HDTV). They have a bit rate of between 600 Mbit/s and 2 Gbit/s.

The storage capacity or the transmission bandwidth available for digital pictures is generally limited. For example, the capacity of a consumer digital video recorder is about 20–40 Mbit/s. To record a CCIR 601 TV signal, the bit rate has to be reduced by a factor of about 4–8, while at the same time a sufficiently high perceptual picture quality has to be maintained. Sometimes even larger reduction factors are required, e.g. for the videophone. The bit rate of telephone channels can be as low as 64–384 kbit/s, so that starting from a CCIR 601 TV signal reduction factors of 400–2500 are required.

Source coding can reduce the bit rate of digital pictures. The theory of source coding has been dealt with in Part I of this book. The chapters which follow will show how the techniques introduced in Part I can be used for reduction of the bit rate of digital pictures.

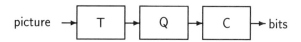

Figure 9.3 *Basic block diagram of source encoder.*

9.4 Picture coding techniques

In Chapter 2 of Part I it was explained that a source-coding system generally consists of a number of different blocks. Figure 9.3 repeats the basic block diagram of a source-coding system. The first block T is a transformation of the signal to some other representation. The transformation is followed by quantization block Q, which limits the accuracy of the output of the transformation. The next block is the coding block C, which removes redundancy from the quantizer output by efficiently mapping the output on to binary codewords. As explained in Chapter 5 in Part I, the transformation is used for two reasons. Firstly, the transformed signal is generally more suited for the removal of irrelevant information. Secondly, the transformation simplifies the removal of redundant information from the quantizer output symbols.

Examples of well-known picture source-coding techniques are given in Chapters 10–14. These techniques are: pulse-code modulation (PCM), differential pulse-code modulation (DPCM), transform coding (TC), subband coding (SBC) and pyramidal coding (PC). Most of these techniques have already been theoretically treated in Part I. As will become clear, these techniques are all named after the type of transformation used on the picture before quantization and coding. The techniques considered in these chapters are all so-called spatial source-coding techniques, since they exploit spatial redundancy in the picture. Sometimes a picture is part of a sequence of pictures, such as digital television signals. Chapter 15 gives an example of a source-coding technique which exploits temporal redundancy in picture sequences by combined coding of pictures.

The description given in Chapters 10–15 is introductory. It is meant to give a first, general overview of the most frequently used picture source-coding techniques. A large number of variations of each technique exists. Discussion of all these variations is beyond the scope of this book. As authors, we have taken the liberty to deal only with those techniques that we consider most illustrative for the principles of picture source coding.

10
Pulse-code modulation of pictures

10.1 Principle

A digital picture is obtained by sampling an analog picture and quantizing each of the samples. Usually, the quantized samples are represented by fixed-length binary codewords. This method of representation is called pulse-code modulation (PCM). In fact, almost all digital pictures have the PCM format before bit-rate reduction is applied. For example, the samples of a CCIR 601 digital TV signal are uniformly quantized with 256 quantization levels, and 8 bits are used to represent the quantized samples.

It is a known phenomenon that the human visual system tolerates a lower quantization accuracy in highly-detailed regions of the picture, such as regions containing edges, than in less detailed ones. This means that highly-detailed regions can be represented with fewer than the original 8 bits per sample. To achieve this lower bit rate, one first has to divide the picture into a number of different regions. Then the amount of detail in each region has to be quantified, and the quantization accuracy has to be made dependent on this amount of detail. This method of source coding is called adaptive PCM (APCM). Figure 10.1 shows a basic block diagram of this procedure.

APCM of one-dimensional signals has been explained in Section 2.2 of Part I. For one-dimensional signals it is usually assumed that they have a mean value equal to zero. APCM can then be performed by splitting the signal into segments of a number of consecutive samples, by measuring the maximum absolute value in each segment, and by adapting the quantization accuracy to this maximum value. This method cannot be directly applied to pictures, since the mean sample value of pictures is generally not equal to zero. The next section gives an example of how this problem can be solved. The system described there is a simplified version of the adaptive dynamic range coding (ADRC) system that has been proposed in [82, 83] for bit-rate reduction of TV signals in consumer magnetic recorders.

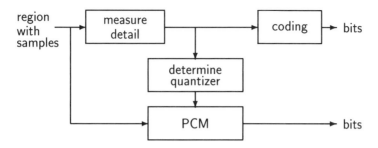

Figure 10.1 *General block diagram of a PCM encoder which adapts the quantization to the amount of detail in a picture.*

10.2 Example of adaptive PCM

Figure 10.2 gives a block diagram of an APCM encoder. In the encoder, the picture is first divided into small, rectangular, non-overlapping regions of, for example, 3×3 or 4×4 samples. This division is illustrated in Figure 10.3. In the remainder of this chapter, the rectangular regions will be called blocks, and the samples values in a block will be indicated as $p[m, n]$, where $m = 1, \ldots, N_r$ is the row index, $n = 1, \ldots, N_c$ is the column index of the sample concerned, and where N_r and N_c are the number of rows and columns in the block, respectively.

For each block, the 'dynamic range' D is calculated, which is the difference between the maximum sample value $p_{\max}$ and the minimum sample value $p_{\min}$ in the block, i.e.

$$D = p_{\max} - p_{\min}. \tag{10.1}$$

The dynamic range D is used as a measure of the amount of detail in the block, and the accuracy with which the samples in the block are quantized depends on the value of the dynamic range. Since the human visual system tolerates more quantization noise in highly-detailed blocks, blocks with a large dynamic range can be more coarsely quantized than blocks with a small dynamic range.

Before quantization, the minimum sample value $p_{\min}$ is subtracted from the samples in the block, i.e. a new block $d[m, n]$ is created, according to

$$d[m, n] = p[m, n] - p_{\min}, \quad \text{for } m = 1, \ldots, N_r; \ n = 1, \ldots, N_c. \tag{10.2}$$

The subtraction of $p_{\min}$ yields a block that is better suited for quantization than the original one. The result of the subtraction is a block that has a minimum value of zero and a maximum value of D, so that the quantizer only has to have representation levels in this range. Both the dynamic range D and the minimum pixel value $p_{\min}$ are transmitted as side information to the decoder.

176 Pulse-code modulation of pictures

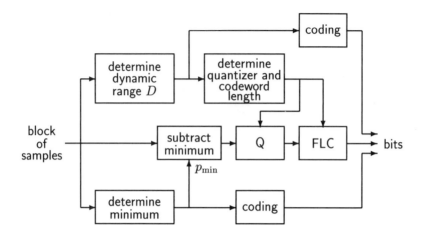

Figure 10.2 *APCM encoder.*

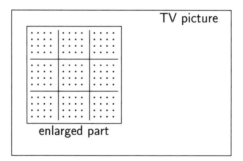

Figure 10.3 *Division of picture into blocks of 4×4 samples.*

The values of $d[m, n]$ are coded with PCM, i.e. they are quantized and subsequently coded with fixed-length coding (FLC). In this example, uniform quantization is assumed. The number of quantization levels, and therefore also the length of the codewords, is determined by the dynamic range. More specifically, the quantization is performed as follows. If the dynamic range is below a certain threshold, only one quantization level is used. In that case, only the dynamic range and the minimum sample value are transmitted, and the decoder reconstructs a block of samples all having a value equal to $(p_{\max} + p_{\min})/2$. For dynamic ranges above the

dynamic range		no. quantization levels $n(D)$	codeword length (in bits)
0 — 3		1	0
4 — 16		2	1
17 — 32		4	2
> 32		8	3

Table 10.1 *Example of bit allocation table.*

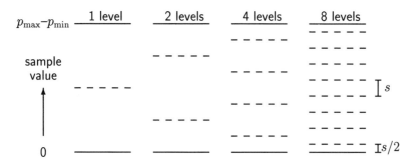

Figure 10.4 *Placement of quantization levels (s is the quantization step size).*

threshold, the quantization step size is determined by

$$\Delta(D) = \frac{D}{n(D)}, \tag{10.3}$$

where $\Delta(D)$ is the step size and $n(D)$ the number of quantization levels used for dynamic range D. An example of the relation between D, $n(D)$ and the codeword length used to represent quantized samples is given in Table 10.1. This table is often called a bit allocation table. Figure 10.4 shows the placements of the quantization levels for $n(D)$ equal to 1, 2, 4 and 8.

The effect of quantizing $d[m,n]$ is the introduction of an additive quantization error $\varepsilon[m,n]$ in the range of $-\Delta(D)/2 \leq \varepsilon[m,n] \leq \Delta(D)/2$, i.e. the reconstructed signal $\hat{d}[m,n]$ is equal to

$$\hat{d}[m,n] = d[m,n] + \varepsilon[m,n]. \tag{10.4}$$

The variance of $\varepsilon[m,n]$ is about $\Delta^2(D)/12$. Actually, this is only true if $\sigma_d \gg \Delta(D)$, where σ_d^2 is the variance of $d[m,n]$.

D and $p_{\min}$ are transmitted to the decoder as side information. The simplest method of coding them is fixed-length coding (FLC). Assuming that R_D bits are used for D, and $R_{\min}$ bits for $p_{\min}$, the bit rate per sample for a block of N_r by N_c samples is equal to

$$R = \frac{N_r N_c \log_2(n(D)) + R_D + R_{\min}}{N_r N_c} \text{ bit/sample.} \qquad (10.5)$$

The total amount of bits needed to represent a complete picture will, of course, depend on the amount of detail in the picture. The smallest rate is achieved for pictures that contain such a low amount of detail that the dynamic ranges of all blocks are below the threshold. Then, assuming 4×4 blocks, and assuming $R_D = R_{\min} = 8$, the minimum bit rate is 1 bit/sample. The maximum rate is achieved for pictures that contain so much detail that all samples are coded with the maximum codeword length. For 4×4 blocks, and for the bit allocation table of Table 10.1, the maximum bit rate is equal to 4 bit/sample. It can be concluded that, in this example, the reduction factor achievable is between 2 and 8. For conventional pictures, such as those of CCIR 601 TV signals, a reduction factor of around 2-3 can be achieved with reasonable picture quality.

In [82, 83] a similar system has been proposed for application in consumer magnetic digital video recorders. Actually, the system described there is more complicated than the one described in this chapter. To increase the coding efficiency, it codes two successive pictures of a video signal in a combined way. Furthermore, it adapts the content of the bit allocation table of Table 10.1 to the amount of detail in a picture, so that a constant bit rate per two pictures is obtained. For details of this system, the reader is referred to the papers concerned.

Figure 10.5 shows the APCM decoder. First, the dynamic range and the minimum are decoded, and they are used to select the quantization levels. Subsequently, the quantized signal $\hat{d}[m,n]$ is decoded (FLD) and reconstructed. A reconstructed block of samples $\hat{p}[m,n]$ is obtained by adding $p_{\min}$ to $\hat{d}[m,n]$.

Figures 10.6, 10.7, and 10.8 show three photographs that illustrate the performance of APCM. Figure 10.6 is the result of coding the picture of Figure 9.2 with the bit allocation given in Table 10.1. The block size was 4×4, and D and $p_{\min}$ were both represented with 8 bits. The average bit rate of the coded picture is 2.71 bit/sample. Figure 10.7 shows the picture before quantization, but after subtraction of the minimum sample values per block. Finally, Figure 10.8 shows the differences between the original and the coded picture.

It could perhaps be argued that the APCM system described here should be classified to the group of predictive coding techniques. One could consider the determination of the minimum sample value $p_{\min}$ as a very rough prediction for all the samples in the block, and the differences $d[m,n]$ between the actual samples and $p_{\min}$ as a prediction error, which is transmitted in quantized form to the decoder. The next chapter will describe a predictive coding technique that attempts to more accurately predict individual sample values.

Example of adaptive PCM   179

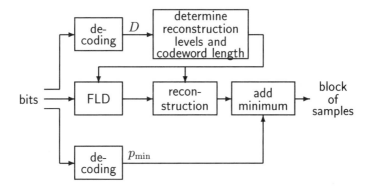

Figure 10.5 *APCM decoder (FLD=fixed-length decoding)*.

Figure 10.6 *Coded picture at 2.71 bit/sample.*

180 Pulse-code modulation of pictures

Figure 10.7 *Picture before quantization, after subtraction of minimum per block. In order to make the samples values better visible, a value of 80 has been added to each sample.*

Figure 10.8 *Differences between original and coded picture, amplified by 6. A value of 128 has been added to map a zero difference to middle grey.*

11
Differential PCM of pictures

11.1 Principle

In the previous chapter it has been explained that PCM quantizes and codes each sample of the picture individually. In Section 5.6 of Part I, it was explained that in general a higher coding efficiency can be obtained with predictive coding. An example of predictive coding is differential pulse-code modulation (DPCM) [84]. In a DPCM system, the difference between a sample value and a prediction of this value is quantized and coded, instead of the sample value itself.

Figure 11.1 shows a DPCM encoder, and Figure 11.2 the corresponding decoder. Briefly summarized, DPCM operates as follows. Assume that the samples of the picture are indicated as $p[m,n]$, with $m = 1, \ldots, M$, $n = 1, \ldots, N$, where M is the number of lines in the picture and N the number of samples per line. A prediction $\tilde{p}[m,n]$ is determined for each sample $p[m,n]$. Subsequently, the difference

$$d[m,n] = p[m,n] - \tilde{p}[m,n], \tag{11.1}$$

is computed. This difference $d[m,n]$, which is often called the prediction error, is quantized and coded.

In the decoder, a reconstructed sample $\hat{p}[m,n]$ is determined by first reconstructing the quantized prediction error $\hat{d}[m,n]$, and subsequently adding the same prediction $\tilde{p}[m,n]$ as has been used in the encoder, i.e.

$$\hat{p}[m,n] = \tilde{p}[m,n] + \hat{d}[m,n]. \tag{11.2}$$

From this equation it is clear that the encoder and decoder must determine exactly the same prediction $\tilde{p}[m,n]$. In the decoder, however, only reconstructed sample values $\hat{p}[m,n]$ are available. The prediction can be derived only from these values. Therefore, also in the encoder the values of $\hat{p}[m,n]$ have to be determined, and the prediction must be determined from these values. This is realized by the prediction method as depicted inside the dashed box in Figure 11.1. This method is called closed-loop prediction.

In Chapter 2 of Part I it has been explained that any source-coding system is composed of a number of different blocks, namely a transformation block, followed

182 Differential PCM of pictures

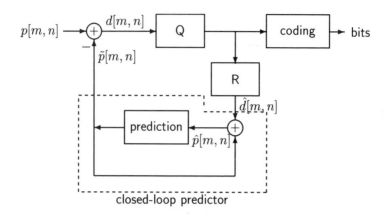

Figure 11.1 *DPCM encoder.*

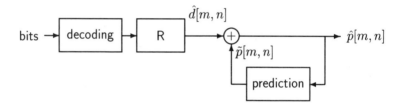

Figure 11.2 *DPCM decoder.*

by a quantization block and a coding block. The subtraction of a prediction from the original samples in a DPCM system can be seen as a transformation of the original samples $p[m,n]$ to a signal $d[m,n]$. This signal $d[m,n]$ has properties such that it is more suited for quantization and coding than $p[m,n]$ itself. As will be illustrated by the example given in Section 11.5, the signal $d[m,n]$ is more suited for the removal of irrelevant information than $p[m,n]$. Furthermore, it will be shown that the quantized prediction error can be efficiently coded with coding techniques of relatively low complexity.

Section 11.2 will introduce a number of commonly used methods for determining the prediction $\hat{p}[m,n]$, and Sections 11.3 and 11.4 will briefly treat quantization and coding of the prediction error.

11.2 Prediction techniques

The predicted sample value $\tilde{p}[m, n]$ has to be determined from previously reconstructed sample values. In order to determine what is meant by 'previously reconstructed', it is, of course, first necessary to define an order in which the samples will be processed. As is usual in picture storage and transmission, it is assumed that pictures are stored or transmitted line by line.

Let us assume that the first $m-1$ lines and the first $n-1$ samples of line m have been processed, which means that the next sample to be treated is sample $p[m, n]$. This is illustrated in Figure 11.3, in which the black dots indicate the already coded samples and the white dots those that still have to be treated.

Usually, linear prediction is performed, i.e. $\tilde{p}[m, n]$ is calculated as a linear combination of reconstructed samples in an environment of $p[m, n]$ [5]. Figure 11.4 shows a commonly used environment, in which the prediction is

$$\hat{p}[m, n] = a \cdot r[m, n-1] + \\ b \cdot r[m-1, n-1] + \\ c \cdot r[m-1, n] + \\ d \cdot r[m-1, n+1], \tag{11.3}$$

where a, b, c and d are the prediction coefficients. To avoid a biased prediction of constant signal values, a, b, c and d are usually chosen such that $a + b + c + d = 1$. A very simple prediction that is sometimes used for picture source coding is the 'previous sample' predictor, for which $a = 1$ and $b = c = d = 0$. In that case, $\tilde{p}[m, n] = \hat{p}[m, n-1]$, which is the value of the previously reconstructed sample on line m.

As will be illustrated in Section 11.5, the prediction error signal $d[m, n]$ usually has a probability of occurrence that is peaked around zero, i.e. small prediction errors occur much more frequently than large ones. This means that $d[m, n]$ can be efficiently represented using variable-length coding. The better the prediction of a sample, the smaller the prediction error, the more peaked the probability, and the smaller the number of bits needed to represent $d[m, n]$.

Without knowledge of the statistics of the picture, it is impossible to determine beforehand which predictor will give the smallest prediction errors. As has been explained in Section 5.6 of Part I, for known statistics it is possible to derive mathematically prediction coefficients which minimize the energy of the quantization error. In general, however, pictures have rapidly changing statistics, which means that the prediction coefficients should be frequently updated and transmitted to the decoder. This is called forward-adaptive prediction. For more information on this topic the reader is referred to [85].

A well-known simple method to cope partially with the varying statistics is that of median prediction [86, 87]. First a number of different predictions $\tilde{p}_k[m, n]$ is computed, where $k = 1, \ldots, K$ and K is an odd integer. For example, for $K = 3$,

184 Differential PCM of pictures

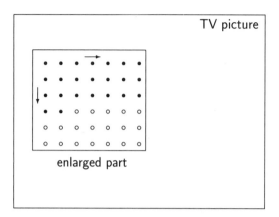

Figure 11.3 *Possible order of treatment of samples in DPCM.*

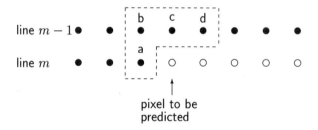

Figure 11.4 *DPCM prediction environment.*

the following predictions could be chosen:

$$\begin{aligned}
\tilde{p}_1[m,n] &= \hat{p}[m, n-1], \\
\tilde{p}_2[m,n] &= \hat{p}[m-1, n], \\
\tilde{p}_3[m,n] &= 0.5\,\hat{p}[m, n-1] + 0.5\,\hat{p}[m-1, n].
\end{aligned} \qquad (11.4)$$

The median value of these three predictions is now chosen as prediction $\tilde{p}[m,n]$, i.e. the predictions are ranked in order of increasing value, and the middle value is chosen. The idea behind median prediction is that the median of a number of predictors will never give the worst prediction, while in some cases it will yield the prediction giving the smallest prediction error. For more details on median prediction the reader is referred to [86, 87].

11.3 Quantization of the prediction error

The prediction error is usually quantized before it is transmitted to the decoder. Let us assume that quantization of the prediction error introduces a quantization error $\varepsilon[m, n]$ such that

$$\hat{d}[m, n] = d[m, n] + \varepsilon[m, n]. \tag{11.5}$$

Substitution of (11.5) into (11.2) gives for the reconstructed picture

$$\begin{aligned} \hat{p}[m, n] &= \tilde{p}[m, n] + \hat{d}[m, n] \\ &= \tilde{p}[m, n] + d[m, n] + \varepsilon[m, n] \\ &= p[m, n] - d[m, n] + d[m, n] + \varepsilon[m, n] \\ &= p[m, n] + \varepsilon[m, n]. \end{aligned} \tag{11.6}$$

It can be concluded that the reconstructed sample value is equal to the original value plus the value of the error introduced by quantization of the prediction error. It is important to note that a quantization error at position $[m, n]$ in the prediction error signal $d[m, n]$ only leads to an error at position $[m, n]$ in the reconstructed picture $\hat{p}[m, n]$. This makes DPCM especially suited for applications in which it is required to accurately control the quantization error in each position of the picture.

An example of such an application is the recording of pictures by what is called an electronic still picture camera (ESP). Instead of storing pictures chemically on photographic film, as is done in conventional cameras, they are stored digitally on, for example, an optical disk. To increase the speed of transfer to the disk, and to increase the storage capacity, bit-rate reduction can be used. The pictures can be retrieved from the disk for printing on photographic paper. Experiments showed that to obtain high-quality prints the picture coding should be of a very high quality. In particular, visible artefacts around edges had to be avoided. DPCM has therefore been proposed [88].

A question that may arise is whether it is possible to indicate more specifically the effect of different quantization characteristics. The answer is yes. Let us, for example, compare the effect of uniform and non-uniform quantization characteristics.

From Section 4.4 of Part I we know that, under the conditions mentioned in that section, uniform quantization of the prediction error signal $d[m, n]$ with a step size equal to Δ will introduce a white-noise quantization error signal $\varepsilon[m, n]$ with a mean value equal to zero and a variance of about $\Delta^2/12$. In the case of uniform quantization, the error signal $\varepsilon[m, n]$ is uncorrelated with the input signal $p[m, n]$.

The effect of non-uniform quantization can be illustrated as follows. As has been explained in Section 4.4 of Part I, a non-uniform quantizer can be thought of as a non-linear function followed by a uniform quantizer. This is illustrated in Figure 11.5. Let us, for example, assume that the non-linear function is equal to $y = f(x) = \sqrt{x}$. Without loss of generality it is assumed that $x \geq 0$. Assuming that the uniform quantization of the signal y of Figure 11.5 introduces a quantization

186 Differential PCM of pictures

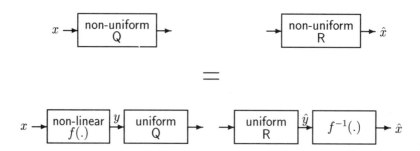

Figure 11.5 *Modelling non-uniform quantization as a non-linear function followed by uniform quantization.*

error ε, it follows that the output signal $\hat{x}$ of the non-uniform reconstruction unit is equal to

$$\hat{x} = (y + \varepsilon)^2 = x + \varepsilon', \tag{11.7}$$

where

$$\varepsilon' = 2\varepsilon\sqrt{x} + \varepsilon^2, \tag{11.8}$$

and $-\Delta/2 \leq \varepsilon \leq \Delta/2$. It can be concluded that the quantization error ε' in the reconstructed signal $\hat{x}$ is a function of the values of the input signal x. In this case, the larger the signal x, the larger the error in the reconstructed signal.

A non-uniform quantizer can be of benefit for the following reason. It is known that the human visual system is less sensitive to noise near very sharp luminance transitions than to noise in areas with a small amount of luminance variations. Advantage can be taken of this phenomenon by applying non-uniform quantization, namely by quantizing large prediction errors more coarsely than small prediction errors. Sharp luminance transitions are generally not well predicted, which results in relatively large prediction error. A coarse quantization of large predictions errors will therefore lead to larger quantization errors near these transitions than in flat areas of the picture. As will be illustrated in the example in Section 11.5, non-uniform quantization can significantly lower the bit rate. The art is to find the quantization characteristic which will introduce artefacts that are just below the threshold of visibility, and at the same time minimizes the bit rate.

11.4 Coding quantized prediction errors

In Section 5.6 of Part I it has been argued that, in the ideal situation, the subtraction of a prediction $\tilde{p}[m,n]$ from the original samples $p[m,n]$, and the subsequent quantization of the prediction error $d[m,n]$ to $\hat{d}[m,n]$, results in a signal of which the first-order entropy is close to the entropy $H_\infty(\hat{d})$. Note that the first-order entropy was defined as $-\sum_i p_i \log_2 p_i$, where p_i is the probability that $d[m,n]$ is quantized to quantization index i. To remove the redundancy of the quantized prediction error signal, relatively simple coding techniques can therefore be used. Very often, each prediction error sample is individually coded with variable-length coding (VLC). An example will be given in the next section.

11.5 Examples of quantization and coding

As an example, Figure 11.6 shows the histogram of the quantized prediction error signal of the picture in Figure 9.2, assuming the median predictor of (11.4) and uniform quantization with a step size equal to 4. The first-order entropy of this signal is equal to 2.83 bit/sample. Table 11.1 shows a Huffman code that is adapted to the statistics shown in Figure 11.6. As can be seen, quantization indices in the range from -32 to 32 are represented with one VLC word. Quantization indices smaller than -32, or larger than 32, are represented by a concatenation of two codewords. The first codeword is a so-called prefix word, which indicates to the decoder that a second codeword will follow. The second codeword is a fixed-length binary word of 9 bits, representing the value of the prediction error. Applying this Huffman code to the picture in Figure 9.2 results in an average bit rate of 2.85 bit/sample, which is almost equal to the first-order entropy.

A lower bit rate at the same perceptual picture quality can be obtained by non-uniform quantization. Table 11.2 gives an example of a possible quantization characteristic. It is assumed that the quantizer is symmetrical about zero, and therefore only the positive part is given. As can be seen, prediction errors around zero are quantized with a step size of 4, whereas errors larger than 46 are quantized with a step size of 20. Figure 11.7 shows the histogram of the quantized prediction error signal. This histogram is somewhat more peaked than the one in Figure 11.1. The first-order entropy is now equal to 2.43 bit/sample, which is 0.4 bit/sample less than the entropy obtained with uniform quantization using a step size of 4. The application of the Huffman code of Table 11.1 results in an average bit rate of 2.54 bit/sample, which is about 10% less than the average bit rate obtained with uniform quantization. The perceptual quality of the pictures coded with both types of quantizers is the same.

As an illustration, Figure 11.8 shows the quantization noise picture obtained with the non-uniform quantization characteristic of Table 11.2. This quantization

188   Differential PCM of pictures

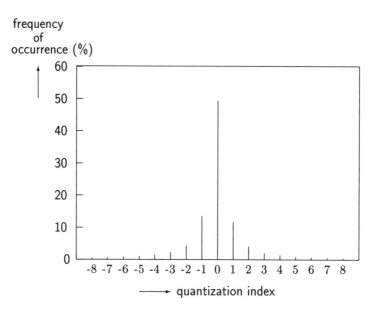

Figure 11.6 *Histogram of quantized prediction error in the case of uniform quantization with a step size of 4.*

noise picture has been constructed by first subtracting the coded picture from the original picture, and then amplifying the result by a factor of 6, after which a value of 128 is added. As can be seen, the quantization noise is clearly correlated with the input picture. The quantization noise picture obtained with uniform quantization contains approximately white noise. It is shown in Figure 11.9.

For CCIR 601 TV pictures such as the one in Figure 9.2, the reduction factor of DPCM is about 2–4 [89]. In general, it is somewhat higher than that of the simple APCM-coding technique described in Chapter 10, but it is lower than that of transform coding and that of subband coding. The latter techniques are described in the next two chapters.

quantization index	codeword	quantization index	codeword
-32	10000010000101	2	10001
-31	10000010000100	3	100100
-30	1001100000111	4	111011
-29	1001100000110	5	1001101
-28	1000001000011	6	1110100
-27	100111111111	7	10000011
-26	111010101000	8	10011101
-25	100101000010	9	11100011
-24	100000100000	10	100000101
-23	10011111110	11	100110001
-22	11100010001	12	100111000
-21	10011001100	13	111010100
-20	10010100000	14	1000000100
-19	1110101011	15	1001010010
-18	1001110011	16	1001010011
-17	1001100001	17	1001100111
-16	1001010001	18	1001111110
-15	1000001001	19	1110001001
-14	111000101	20	10000010001
-13	100111110	21	10011000000
-12	100110010	22	10011100100
-11	100000011	23	11100010000
-10	11101011	24	11101010101
-9	10011110	25	100110000010
-8	10010101	26	100101000011
-7	10000000	27	100110011010
-6	1110000	28	100111111110
-5	1001011	29	100111001010
-4	111001	30	111010101001
-3	100001	31	100110011011
-2	1111	32	100111001011
-1	101	>32 or <-32	1000000101 as prefix + 9-bit codeword
0	0		
1	110		

Table 11.1 *Huffman code for quantized prediction errors.*

190 Differential PCM of pictures

quantization level	decision level
0	2
4	7
11	16
22	29
37	46
56	66
76	86
96	106
116	126
136	146
156	166
176	186
196	206
216	226
236	255

Table 11.2 *Possible non-uniform symmetrical quantization characteristic (only positive part is given).*

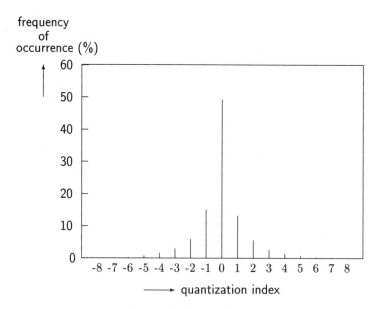

Figure 11.7 *Histogram of quantized prediction error using non-uniform quantization.*

Examples of quantization and coding 191

Figure 11.8 *Difference picture in case of non-uniform quantization (times 6, plus 128)*.

Figure 11.9 *Difference picture in case of uniform quantization (times 6, plus 128)*.

12

Transform coding of pictures

12.1 Introduction

At present the most widely used technique for source coding of pictures is block-based transform coding (TC) [90]. Figure 12.1 shows a block diagram of a TC system. The picture is first divided into small, rectangular, non-overlapping blocks of, for example, 8×8 or 16×16 samples. Each block of samples is transformed to a transform domain, resulting in a set of transform coefficients. Usually, the transform is chosen such that its transform coefficients are related to spatial frequencies in the picture. Section 12.2 will introduce the most popular transform, namely the discrete cosine transform (DCT). The transform coefficients are quantized and subsequently coded to binary codewords. The coded coefficients are transmitted to the decoder. After decoding, the quantized transform coefficients are reconstructed and inverse-transformed to a block of reconstructed samples.

The high-frequency coefficients are usually quantized more coarsely than the low-frequency ones. This can be allowed since the human visual system is less sensitive to noise in the higher spatial frequencies than to noise in the lower frequencies. Section 12.3 discusses quantization in greater detail.

The quantized transform coefficients generally contain a significant amount of redundancy. They are therefore often represented with variable-length coding. Section 12.4 deals with coding of the quantized coefficients.

The use of variable-length coding results in a variable number of bits per block. Most applications, however, require a constant bit rate. To ensure that transform coefficients can be transmitted at a constant rate, some bit-rate control mechanism has to be present. As has been mentioned in Section 6.10 of Part I, basically two bit-rate control techniques can be used, namely forward adaptive bit-rate control, and backward adaptive bit-rate control. The latter is often indicated as buffer regulation or buffer control. In transform-coding systems this is the technique most frequently used.

Section 12.5, finally, describes a number of specific applications of TC.

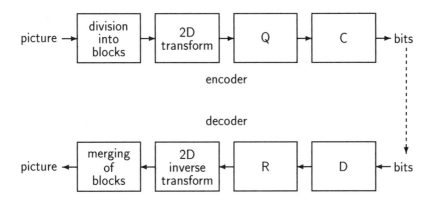

Figure 12.1 *General block diagram of a transform-coding system (without bit-rate control).*

12.2 The transform

Let us assume that the samples in the block to be coded are denoted by $p[m,n]$, with $m = 0, \ldots, M-1$ and $n = 0, \ldots, N-1$, with M the number of lines in the block and N the number of samples per line, respectively. The block of samples can be considered to be an $M \times N$ matrix:

$$\mathbf{P} = \begin{pmatrix} p[0,0] & p[0,1] & \cdots & p[0, N-1] \\ p[1,0] & p[1,1] & \cdots & p[1, N-1] \\ \vdots & \vdots & \cdots & \vdots \\ p[M-1,0] & p[M-1,1] & \cdots & p[M-1, N-1] \end{pmatrix}. \quad (12.1)$$

Section 5.4 of Part I has already indicated how the transformation is performed. Two mathematical formulations of the transformation have been given there. In the first formulation, in Subsection 5.4.1, the transformation has been written as pre- and post-multiplication of the matrix $\mathbf{P}$ with transform matrices $\mathbf{M}_v$ and $\mathbf{M}_h$, i.e.

$$\mathbf{C} = \mathbf{M}_v \mathbf{P} \mathbf{M}_h, \quad (12.2)$$

and

$$\mathbf{P} = \mathbf{M}_h \mathbf{C} \mathbf{M}_v, \quad (12.3)$$

where

$$\mathbf{C} = \begin{pmatrix} c[0,0] & c[0,1] & \cdots & c[0, N-1] \\ c[1,0] & c[1,1] & \cdots & c[1, N-1] \\ \vdots & \vdots & \cdots & \vdots \\ c[M-1,0] & c[M-1,1] & \cdots & c[M-1, N-1] \end{pmatrix}. \quad (12.4)$$

is the matrix of transform coefficients.

In the second formulation, given in Subsection 5.4.5, the transformation has been written as the decomposition of the matrix $\mathbf{P}$ into a linear combination of basis matrices, i.e.

$$\mathbf{P} = c[0,0]\mathbf{B}_{00} + c[1,0]\mathbf{B}_{10} + \ldots + c[M-1, N-1]\mathbf{B}_{M-1,N-1}$$

$$= \sum_{i=0}^{M-1} \sum_{j=0}^{N-1} c[i,j]\mathbf{B}_{ij}. \qquad (12.5)$$

In this expression the matrices $\mathbf{B}_{ij}$, with $i = 0, \ldots, M-1$ and $j = 0, \ldots, N-1$, indicate the $M \times N$ different basis matrices, and $c[i,j]$ are the transform coefficients.

The question is how to choose the set of basis matrices $\mathbf{B}_{ij}$. To determine a suitable set a number of aspects have to be considered. First, as explained in Subsection 5.4.2 of Part I, one would like the transform to concentrate the signal energy into a limited number of preferably mutually uncorrelated transform coefficients. This simplifies the removal of redundant information from the coefficients since relatively simple coding techniques can then be used. Second, as explained in 5.4.5, one would want the transform coefficients to be perceptually relevant, i.e. the basis matrices $\mathbf{B}_{ij}$ should, for example, be related to spatial frequencies. In that case, the quantization accuracy applied to every transform coefficient can be optimally adapted to the properties of our human visual system. Third, it would be desirable to use transforms that can be implemented with a low complexity.

Since the properties of the human visual system are not yet fully known, the transforms that are most frequently used are those that result in uncorrelated or nearly uncorrelated transform coefficients. The best-known transformation at the moment is the discrete cosine transform (DCT). A number of definitions of the DCT exist. The one that is usually used is [91]

$$c[i,j] = \sqrt{\frac{2}{M}} \sqrt{\frac{2}{N}} K[i,j] \cdot \sum_{m=0}^{M-1} \sum_{n=0}^{N-1} p[m,n] \cdot \cos\left(\frac{(2m+1)i\pi}{2M}\right) \cos\left(\frac{(2n+1)j\pi}{2N}\right), \quad (12.6)$$

where

$$K[i,j] = \begin{cases} \frac{1}{2} & \text{for } i = j = 0 \\ \frac{1}{\sqrt{2}} & \text{for } i = 0, j \neq 0, \text{ or } \\ & \quad i \neq 0, j = 0 \\ 1 & \text{for } i \neq 0 \text{ and } j \neq 0. \end{cases} \qquad (12.7)$$

The inverse DCT is then defined as

$$p[m,n] = \sqrt{\frac{2}{M}} \sqrt{\frac{2}{N}} \sum_{i=0}^{M-1} \sum_{j=0}^{N-1} K[i,j]c[i,j] \cdot \cos\left(\frac{(2m+1)i\pi}{2M}\right) \cos\left(\frac{(2n+1)j\pi}{2N}\right). \quad (12.8)$$

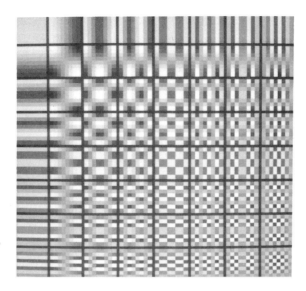

Figure 12.2 *Basis matrices of an 8×8 DCT.*

Figure 12.2 shows the set of 64 basis matrices of the DCT, assuming transformation of blocks of 8×8 samples.

The popularity of the DCT stems from a number of its properties. Firstly, its transform coefficients usually have a low correlation. Relatively low-complexity coding techniques can therefore be used to remove redundant information from quantized DCT coefficients.

Secondly, the two-dimensional DCT is a separable transform, i.e. it can be calculated by first transforming each row of the block by a one-dimensional DCT and subsequently transforming each column of the resulting block by a one-dimensional DCT. The one-dimensional DCT of a signal $x[n]$, with $n = 0, \ldots, N-1$, is defined as

$$c[k] = \sqrt{\frac{2}{N}} K[k] \sum_{n=0}^{N-1} x[n] \cos\left(\frac{(2n+1)k\pi}{2N}\right), \qquad (12.9)$$

for $k = 0, \ldots, N-1$, where $K[0] = 1/\sqrt{2}$, and $k[k] = 1$ for $k \neq 0$. There exist a number of methods for fast calculation of the DCT [92].

A third reason why the DCT is popular is that its basis matrices are closely related to spatial frequencies. This can be understood as follows. The one-dimensional DCT of a row of N samples can be calculated by calculating the discrete Fourier transform (DFT) of a signal $y[n]$, with

$$\begin{aligned} y[n] &= x[n] & \text{for} \quad n = 0, \ldots, N-1 \\ &= x[2N-1-n] & \text{for} \quad n = N, \ldots, 2N-1. \end{aligned} \qquad (12.10)$$

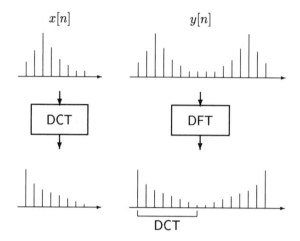

Figure 12.3 *Illustration of calculation of the DCT directly and via the DFT.*

The DFT of $y[n]$ is equal to

$$Y[k] = \sum_{n=0}^{2N-1} y[n] e^{-j(2\pi kn/2N)},$$
$$= 2e^{j(k\pi/2N)} \sum_{n=0}^{N-1} y[n] \cos\left(\frac{(2n+1)k\pi}{2N}\right). \quad (12.11)$$

The DCT can then be calculated from

$$c[k] = \frac{1}{\sqrt{2N}} K[k] e^{-j(k\pi/2N)} Y[k], \quad k = 0, \ldots, N-1. \quad (12.12)$$

Figure 12.3 shows the relation between $x[n]$ and $y[n]$ graphically. As can be seen, for $n = N, \ldots, 2N-1$, $y[n]$ is obtained by mirroring $x[n]$. The first N terms of the discrete Fourier transform of $y[n]$ are used to determine the DCT coefficients. Shaping of the power spectral density of the quantization noise is now possible by varying the accuracy with which different transform coefficients are quantized. This will be further discussed in the next section.

12.3 Quantizing of transform coefficients

This section deals with the quantization of transform coefficients. First, in Subsection 12.3.1, the effect of the quantization on the reconstructed picture will be analysed. Then, in Subsection 12.3.2, it will be explained how, by so-called weighting of the transform coefficients, a frequency-dependent quantization error can be introduced. Finally, in Subsection 12.3.3, the choice of quantizer characteristics will be discussed.

12.3.1 The effect of quantization

We assume that the reconstructed transform coefficients in the decoder are denoted by $\hat{c}[i,j]$. These reconstructed coefficients are quantized versions of the original coefficients $c[i,j]$, i.e.

$$\hat{c}[i,j] = Q(c[i,j]), \tag{12.13}$$

where the function Q is the quantization characteristic. Assume that quantization of transform coefficient $c[i,j]$ to $\hat{c}[i,j]$ results in an additive quantization error $\varepsilon[i,j]$, i.e.

$$\hat{c}[i,j] = c[i,j] + \varepsilon[i,j]. \tag{12.14}$$

From (12.5) it follows that the reconstructed samples are equal to

$$\hat{\mathbf{P}} = \sum_{i=0}^{M-1} \sum_{j=0}^{N-1} \hat{c}[i,j] \mathbf{B}_{ij}, \tag{12.15}$$

where $\hat{\mathbf{P}}$ is an $M \times N$ matrix with elements $\hat{p}[m,n]$, and $\hat{p}[m,n]$ is the nth reconstructed sample on the mth row in the block of samples. From substitution of (12.14) into (12.15) it follows that

$$\hat{\mathbf{P}} = \mathbf{P} + \mathbf{E}, \tag{12.16}$$

where

$$\mathbf{E} = \sum_{i=0}^{M-1} \sum_{j=0}^{N-1} \varepsilon[i,j] \mathbf{B}_{ij} \tag{12.17}$$

is the matrix of reconstruction errors, with elements $e[m,n]$, where $e[m,n]$ is the reconstruction errors of the sample at position $[m,n]$ in the block, i.e.

$$\hat{p}[m,n] = p[m,n] + e[m,n]. \tag{12.18}$$

Assuming that $b_{ij}[m,n]$ indicates the element of basis matrix $\mathbf{B}_{ij}$ on the mth row and the nth column, it follows from (12.17) that

$$\begin{aligned} e[m,n] &= \varepsilon[0,0]\, b_{00}[m,n] + \varepsilon[1,0]\, b_{10}[m,n] + \ldots \\ &\quad + \varepsilon[M-1, N-1]\, b_{M-1,N-1}[m,n] \\ &= \sum_{i=0}^{N-1} \sum_{j=0}^{M-1} \varepsilon[i,j] b_{ij}[m,n]. \end{aligned} \tag{12.19}$$

Figure 12.4 *Left: original block of 8×8 samples, right: corresponding DCT coefficients.*

This expression shows that the reconstruction error at position $[m, n]$ is equal to a weighted sum of the basis matrices at position $[m, n]$. The weights are equal to the errors made by quantizing the transform coefficients. For example, DCT basis matrices contain two-dimensional, cosine-like patterns. A large quantization error in one of the coefficients is visible as a cosine-like pattern added to the original block of samples. This is illustrated in Figures 12.4 and 12.5. Figure 12.4 shows on the left an uncoded 8×8 block of samples containing a diagonally-oriented edge, and on the right the magnitudes of the corresponding DCT coefficients. Figure 12.5 shows on the right the DCT coefficients with coefficient $c[2, 1]$ quantized to zero, and on the left the resulting block of samples after inverse transformation.

12.3.2 Frequency-dependent quantization

It is a well-known phenomenon that for transforms like the DCT our human visual system is less sensitive to quantization errors in the high-frequency transform coefficients than to those in the low-frequency ones. This means that it is permissible to quantize the high-frequency transform coefficients more coarsely than the low-frequency ones. This lowers the bit rate needed to represent the quantized coefficients.

Quantizing transform coefficients with different accuracies is often realized by multiplying each coefficient $c[i, j]$ by a weighting factor $w[i, j]$ before quantization.

Quantizing of transform coefficients 199

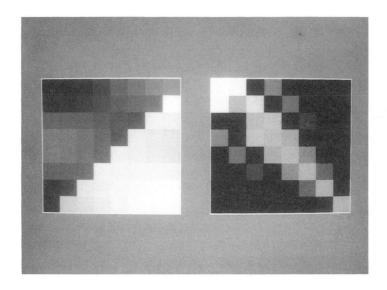

Figure 12.5 *Left: block of samples after inverse DCT, right: block of DCT coefficients with c[2,1]=0.*

i.e. instead of quantizing coefficient $c[i,j]$, a coefficient

$$c_w[i,j] = w[i,j] \cdot c[i,j] \tag{12.20}$$

is quantized. All weighted transform coefficients $c_w[i,j]$ are then quantized with the same quantization characteristic. Assume again that the quantization introduces a quantization error $\varepsilon[i,j]$, i.e.

$$\hat{c}_w[i,j] = c_w[i,j] + \varepsilon[i,j], \tag{12.21}$$

where $\hat{c}_w[i,j]$ is the quantized version of coefficient $c_w[i,j]$ in the decoder. Before inverse transformation, inverse weighting is applied to reconstruct the quantized coefficient $\hat{c}[i,j]$, i.e.

$$\hat{c}[i,j] = \frac{\hat{c}_w[i,j]}{w[i,j]}. \tag{12.22}$$

Figure 12.6 illustrates the weighting procedure. Combination of (12.20), (12.21) and (12.22) gives, for the quantized coefficients,

$$\hat{c}[i,j] = c[i,j] + \frac{\varepsilon[i,j]}{w[i,j]}. \tag{12.23}$$

From (12.23) it can be seen that the quantization error can be made large by choosing small values for $w[i,j]$ and vice versa. Usually, the $w[i,j]$ are chosen such

200   Transform coding of pictures

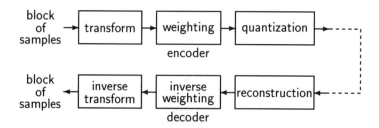

Figure 12.6 *Block diagram of transform coding with weighting.*

```
1.00  1.00  0.90  0.84  0.74  0.68  0.58  0.52
1.00  0.97  0.90  0.81  0.74  0.65  0.58  0.48
0.90  0.90  0.84  0.77  0.71  0.65  0.55  0.48
0.84  0.81  0.77  0.74  0.68  0.61  0.52  0.45
0.74  0.74  0.71  0.68  0.61  0.55  0.48  0.42
0.68  0.65  0.65  0.61  0.55  0.48  0.42  0.35
0.58  0.58  0.55  0.52  0.48  0.42  0.39  0.32
0.52  0.48  0.48  0.45  0.42  0.35  0.32  0.26
```

Figure 12.7 *Example of a weighting function for DCT coding.*

that $0 < w[i,j] \leq 1$. Figure 12.7 shows an example of a set of weighting factors. Such a set is usually called a weighting function.

The entropy of quantized weighted coefficients is generally much lower than that of quantized unweighted coefficients. This means that weighting leads to a substantial reduction of the bit rate. Assume that, for example, the picture of Figure 9.2 is transformed with the DCT, using blocks of 8×8 samples. Uniform midtread quantization of the unweighted DCT coefficients with step size 4 results in quantized coefficients with an average first-order entropy of 2.98 bit/coefficient. Applying the same quantization characteristic to coefficients weighted with the weighting function of Figure 12.7 gives an entropy of 2.37 bit/coefficient. The art is to find the weighting function that gives the largest decrease in bit rate, while not decreasing the perceptual picture quality. Determination of weighting functions is generally performed by perceptual experiments.

12.3.3   Quantization characteristics

The previous sections have dealt with the effect of quantization of a transform coefficient on the reconstructed picture, and with the question of how to obtain a frequency-dependent quantization. This section deals with the question of which quantization characteristic should be used.

First, it has to be determined what type of quantization should be applied. In principle, all quantization techniques treated in Chapter 4 of Part I are applicable. A factor that can determine the type is the allowed complexity of the transform-coding system. If a relatively low complexity is required, straightforward uniform quantization can be used. If a higher complexity is tolerable, the theoretically optimal technique of vector quantization may be applied [93, 94]. Another factor that can determine the type of quantization is the technique used to code quantized transform coefficients. Suppose that, for example, each quantized coefficient is coded separately. Then it is known from Section 4.4.3 of Part I that, assuming the mean-square error as the quality criterion, the technique of Max-Lloyd quantization is the optimal solution if we apply fixed-length coding. For variable-length coding uniform quantization is optimal. This only holds for quantizers with a sufficiently large number of quantization levels.

Once the type of quantization has been chosen, it has to be determined how accurately each coefficient has to be quantized, i.e. the number of quantization levels per coefficient has to be chosen. This is sometimes referred to as bit allocation. Furthermore, the position of the quantization levels has to be determined. The optimal bit allocation is the one that, given a bit rate, divides the available bits such that the amount of quantization noise is minimized. As will be illustrated by the following example, the optimal bit allocation depends both on the signal statistics and on the quality criterion used for comparing original and coded pictures.

Assume that the statistics of the picture are known, and that they do not vary over the picture. Let σ_{ij}^2 be the variance of coefficient $c[i,j]$, with $i = 0, \ldots, M-1$, and $j = 0, \ldots, N-1$. Assume further that the picture has to be coded at a rate of R bit/sample, and that each coefficient is individually quantized with Max-Lloyd quantization, and coded with fixed-length coding. The quantization will introduce a certain amount of quantization noise. Let $\sigma_{q,ij}^2$ denote the variance of the quantization noise added to coefficient $c[i,j]$. The total amount of quantization noise is equal to

$$\sigma_q^2 = \sum_{i=0}^{M-1} \sum_{j=0}^{N-1} \sigma_{q,ij}^2. \tag{12.24}$$

Assume that all transform coefficients are considered to be equally important. The total amount of quantization noise σ_q^2 can then be seen as a measure of the quality of the coded picture. The lower σ_q^2, the higher the picture quality. The question is how many quantization levels L_{ij} should be used for coefficient $c[i,j]$ in order to minimize σ_q^2. In, for example, [4] it is explained that the optimal number of bits for coefficient $c[i,j]$ is given by

$$R_{ij} = R + \frac{1}{2} \log_2 \left(\frac{\sigma_{ij}^2}{\left[\prod_{k=0}^{M-1} \prod_{l=0}^{N-1} \sigma_{kl}^2 \right]^{1/MN}} \right). \qquad (12.25)$$

The number of levels for coefficients $c[i,j]$ is then given by

$$L_{ij} = 2^{R_{ij}}. \qquad (12.26)$$

Note that rounding of L_{ij} may be required to obtain an integer number.

Figure 12.8(b) shows an example of a bit allocation for a block of 8×8 transform coefficients. In this figure, the upper left coefficient represents the lowest horizontal and vertical spatial frequency. In this example, the lower frequencies are represented at a much higher bit rate than the higher ones. To some high-frequency coefficients no bits are allocated at all. These coefficients are not transmitted; in the decoder, they are put at zero. The above-described method of bit allocation is often indicated as zonal sampling, or zonal coding, since only a predefined zone of coefficients is quantized and coded [95]. This is illustrated schematically in Figure 12.8(a).

In the example given above it has been assumed that all coefficients are equally important. In that case, the effect of quantization of a coefficient on the picture quality is independent of the place of this coefficient in the block. As has been explained in Subsection 12.3.2, however, high-frequency coefficients can generally be quantized more coarsely than low-frequency ones. The bit allocation explained above can be slightly modified to achieve this frequency-dependent quantization by replacing σ_{ij} in (12.25) by $w[i,j] \cdot \sigma_{ij}$, where $w[i,j]$ is the weighting factor for coefficient $c[i,j]$ from Subsection 12.3.2, and σ_{kl} by $w[k,l] \cdot \sigma_{kl}$. This is similar to applying the bit allocation according to (12.25) to the weighted transform coefficients.

Up to now it has been assumed that the statistics of the picture are constant. In reality this is rarely true. Normal pictures consist of areas with a large amount of detail, such as edges and texture, and areas with a low amount of detail, i.e. areas in which the luminance is almost constant. Blocks from the low-detailed areas have only a small number of non-zero transform coefficients. Application of the bit allocation of Figure 12.8 would mean that a considerable quantity of bits is spent on zero-valued coefficients.

Zonal coding can be made more efficient by making the selection of the zone adaptive to the contents of the block of samples. Figure 12.9 shows schematically a possible realization of adaptive zonal coding. Before quantization and coding, the block of transform coefficients is classified in one of a number of classes. This classification can be performed, for example, on the basis of the energy in the block of samples. Then, a different bit allocation table is used for each class. For low-energy blocks, only a small number of coefficients is coded, whereas for high-energy blocks almost all are coded. It is assumed that the decoder has the same set of bit

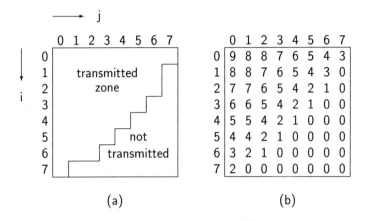

Figure 12.8 (a) Zonal coding using a fixed zone, (b) example of bit allocation.

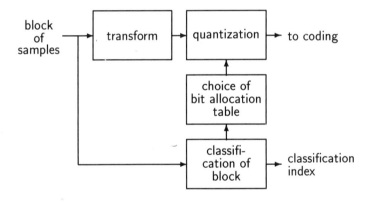

Figure 12.9 Adaptive zonal coding.

allocation tables as the encoder. Information about the class of the block is sent to the decoder, so that it can select the proper bit allocation table. For a detailed example of such a system see [95].

A method that has proved to be more effective than zonal sampling is threshold sampling or, as it is usually called, threshold coding [96]. In a threshold coding system, only coefficients with a value above a certain threshold are quantized, coded and transmitted. Of course, apart from the values of these coefficients, their position within the block has also to be transmitted to the decoder. The position is often called the address. An example of threshold coding is given in the next section.

The advantage of threshold coding is that it automatically adapts to the contents

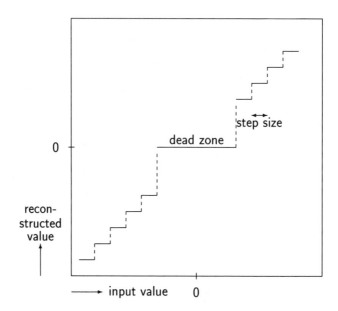

Figure 12.10 *Characteristic of a dead-zone quantizer.*

of a block. For low-energy blocks only a small number of coefficients is above the threshold. Only a small number of bits is needed to transmit these coefficients. For high-energy blocks a large number of coefficients is transmitted. As a result, the bit rate per block depends on the coefficients. This implies that the bit rate per picture depends on the contents of a picture. Usually, a constant bit rate per picture, or per part of a picture, is required. In order to achieve this, threshold coding is usually combined with a feedback buffer regulation mechanism. This mechanism has been explained in a general way in Section 6.10 of Part I.

Thresholding and quantizing transform coefficients are usually performed in combination. Assume, for example, that uniform quantization of the coefficients above the threshold is performed. Combined thresholding and quantization can be performed with the quantization characteristic of Figure 12.10 [97]. This method of quantization is often called dead-band or dead-zone uniform quantization. As can be seen in Figure 12.10, there is a certain 'dead-band' around zero. All input values falling in this band are quantized to zero, while those outside the band are uniformly quantized. The advantage of a dead-band uniform quantizer in comparison with a pure uniform quantizer is that it significantly decreases the entropy of the quantized coefficients, so that a lower bit rate can be achieved. If the dead band is sufficiently small, the somewhat coarser quantization of coefficients with a very small value does not introduce any significant distortion into the reconstructed picture.

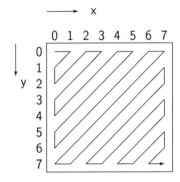

Figure 12.11 *Zig-zag scanning of coefficients.*

12.4 Coding of quantized coefficients

The quantized transform coefficients have to be mapped on to binary codewords before they can be stored or transmitted. This section gives an example of how this can be performed.

At present, threshold coding is the most widely used technique for selecting which coefficients should be quantized and coded. It is usually combined with variable-length coding of the quantized coefficients, which generally contain a significant amount of redundancy. Examples of the statistics of transform coefficients can be found in [98].

The following will give a simple example of a combination of variable-length coding of non-zero transform coefficients and runlength coding of zero-valued coefficients. In this example it is assumed that the quantized transform coefficients can only have the value 0, 1, 2 or 3. Of course, the range of possible values is much larger in reality.

The first step is to scan the block of coefficients in a certain order. A frequently used scanning pattern is the so-called zig-zag scanning pattern, which is indicated in Figure 12.11. Figure 12.12(a) shows a block of 8×8 quantized transform coefficients. After scanning with the zig-zag method of Figure 12.11, a row of coefficients as given in Figure 12.12(b) results. The values of the non-zero coefficients, generally referred to as the amplitudes, and their positions have to be transmitted. This can be done by coding the amplitudes with variable-length codewords, and coding strings of zero-valued coefficients with runlength coding.

Table 12.1 shows the codewords used to represent the non-zero amplitudes and the strings of zeros. During coding, it is assumed that each non-zero coefficient is followed by another non-zero coefficient. If this is not the case, it has to be specifically indicated. Thereto, the amplitude VLC table contains a special code-

```
            3 2 3 2 2 0 1 0
            2 2 2 2 0 0 0 0
            1 2 1 0 0 0 0 0
    (a)     1 1 0 0 0 0 0 0
            0 1 1 0 0 0 0 0
            1 1 0 0 0 0 0 0
            1 0 0 0 0 0 0 0
            0 0 0 0 0 0 0 0
```

(b)   32212322210112200001111100010 ... 0

Figure 12.12 *(a) Block of quantized coefficients, and (b) result of zig-zag scanning of this block.*

word. This codeword, which is called the runlength-coding (RLC) prefix, is placed between a non-zero coefficient and a string of zeros. Note that a string of zeros is always followed by a non-zero coefficient. The amplitude VLC table further contains a special codeword which is placed after the last coded non-zero coefficient. This codeword is often called the end-of-block (EOB) word.

A string of zeros is coded using so-called runlength coding (RLC), i.e. it is represented by a variable-length codeword. In Table 12.1(b), only strings up to a length of 6 are translated to one codeword. For strings larger than 6, a prefix is first generated, which is followed by a fixed-length codeword expressing the length of the string. Since the maximum length can be 64, the length of this fixed-length codeword has to be 6 bits.

Figure 12.13 shows how the row of coefficients is mapped on to a stream of bits. In this simple example, 55 bits are used to represent the block of 8×8 samples, which is a bit rate of 0.86 bit/sample.

12.5 Specific applications

The Joint Picture Experts Group (JPEG) of the International Standardization Organization (ISO) has recently proposed a standard for the storage and transmission of still pictures. This method is based on transform coding, with the discrete cosine transform as transform. Interested readers can find a description of the system in [99].

The technique of transform coding has also been proposed for the magnetic

amplitude	codeword
1	0
2	10
3	110
EOB	1110
RLC prefix	1111

(a)

runlength	codeword
1	0
2	100
3	101
4	110
5	1110
6	11110
prefix	11111

(b)

Table 12.1 *(a) Amplitude VLC table, and (b) runlength VLC table.*

recording of digital TV signals. Information about this topic can be found in, for example, [100, 101, 102]. With the system mentioned in [100], the bit rate of CCIR 601 TV signals can be reduced from 166 to 20–30 Mbit/s, while maintaining a high picture quality. Of special interest in this paper is the way in which the interlacing of the CCIR 601 signal is dealt with. The following is a brief description of how it operates.

In order to understand the system, it is first necessary to understand interlacing. In Chapter 9 it has be explained that digital pictures are obtained by sampling a projection of the luminance radiated or reflected by objects on a flat, rectangular area, and by quantizing each sample. In TV cameras, sampling is usually performed by scanning the projection as indicated in Figure 12.14. At a certain instant, the odd lines are scanned, sampled and quantized. At a later instant, the even lines are processed. The odd and even lines are combined to obtain a single TV picture. This TV picture is usually called a TV frame. The sets of odd and even lines are called fields. For example, a TV signal with 25 pictures per second has 50 fields per second, and the time interval between two fields is equal to 20 msec.

The system proposed in [100] is called motion-adaptive intraframe DCT coding. Figure 12.15 shows its block diagram. It operates on frames of the TV signal. First,

208 Transform coding of pictures

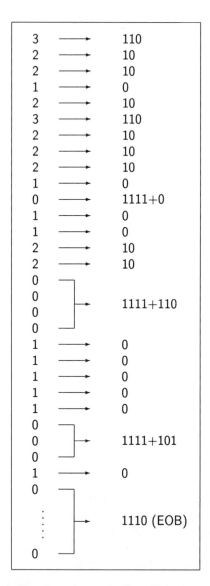

Figure 12.13 *Mapping of quantized coefficients to codewords.*

each frame is divided into blocks of 8×8 samples. For each block, it is determined whether it contains moving objects. Blocks without moving objects are transformed with an 8×8 DCT. Blocks with moving objects are first divided into two blocks of 4×8 samples, which are both transformed with a 4×8 DCT. Figure 12.16 shows

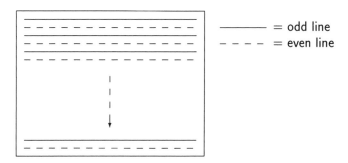

Figure 12.14 *Scanning of an interlaced TV picture: first the odd and then the even lines are scanned.*

how this division is performed. The first block only contains samples of the first field of the interlaced TV frame, whereas the second block only contains samples of the second field.

The division of blocks with moving objects has the following advantage. Interlacing leads to the generation of high vertical spatial frequencies in those parts of the TV frame that contain moving objects. Transformation of a block of 8×8 samples with motion would result in a large number of non-zero transform coefficients. In order to preserve the motion, these coefficients have to be accurately quantized. Splitting the block before transformation, however, results in a significantly lower number of non-zero transform coefficients. The two blocks of 4×8 samples can therefore be more efficiently represented, with about 20 to 30% fewer bits than the block of 8×8 samples.

The block of 8×8 transform coefficients, or the two blocks of 4×8 transform coefficients, are quantized and coded with variable-length coding. For a specific description of how this is performed the reader is referred to [100].

210   Transform coding of pictures

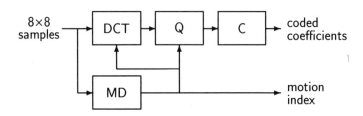

Figure 12.15 *Motion-adaptive intraframe DCT coding: the encoder. MD=motion detection, Q=quantization, C=coding.*

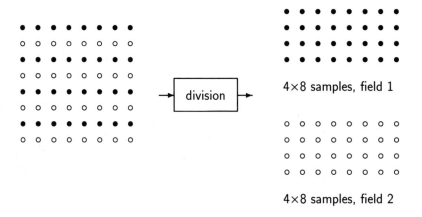

Figure 12.16 *Division of a block of 8×8 samples into two blocks of 4×8 samples.*

13
Subband coding of pictures

13.1 Principle

This chapter deals with subband coding (SBC) of pictures [103]. For simplicity, subband coding will first be explained for one-dimensional digital signals. Then the generalization to two-dimensional signals such as pictures will be made.

Figure 13.1 repeats the general block diagram of a subband coding system of Section 5.5 of Part I. The input signal $x[n]$ is split into N different frequency bands in the encoder. This is usually done with a filter bank consisting of one lowpass filter, $N-2$ bandpass filters, and one highpass filter. The signals resulting from the splitting have a limited bandwidth. It is therefore permissible to subsample them. In the scheme of Figure 13.1 it is assumed that all bands have an equal bandwidth of $1/N$ times the bandwidth of the input signal $x[n]$. It is then permissible to subsample them with a factor N, i.e. only every Nth sample is preserved. Section 13.2 describes a commonly used method for realization of the filter bank.

The subsampled signals, henceforth indicated as subband signals, are quantized and coded. As is the case for transform coding, they can be quantized with different accuracies. This makes it possible to adapt the power spectral density of the quantization noise to the characteristics of the human visual system. Quantization and coding of subband signals are discussed in Section 13.3.

In the decoder, the coded subband signals are decoded, reconstructed, interpolated and added to form a reconstructed signal $\hat{x}[n]$. The interpolation is performed as follows. The reconstructed subband signals are first upsampled by a factor N, i.e. $N-1$ zero-valued samples are placed between each two samples of each reconstructed subband signal. The upsampled signals are then filtered using one lowpass and $N-1$ bandpass filters.

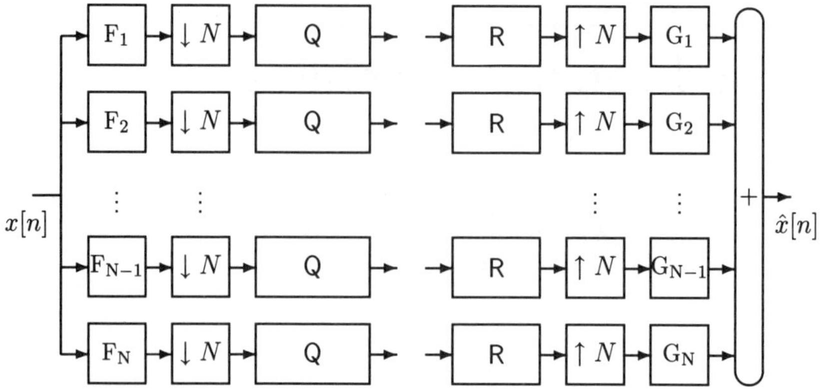

212 Subband coding of pictures

Figure 13.1 *Basic scheme of a subband source encoder and decoder.*

13.2 The filter bank

13.2.1 Filtering of one-dimensional signals

Figure 13.2 illustrates the splitting and reconstruction procedure for a simple subband coding system consisting of only two subbands with the same bandwidth. We denote the impulse responses of the lowpass and highpass filters in the splitting unit as $h_l[n]$ and $h_h[n]$, respectively, and those in the reconstruction unit as $g_l[n]$ and $g_h[n]$. Let us first assume that the lowpass and highpass filters are ideal filters, i.e. that they have unity gain in the passband, and zero gain in the stopband, and that the subband signals are not quantized. Under these assumptions, the output signal $\hat{x}[n]$ is exactly equal to the input signal $x[n]$. In practical situations, however, the lowpass and highpass filters are not ideal, so that they do not completely remove the energy in the stopband. As will be explained below, this may cause the output signal $\hat{x}[n]$ not to be equal to the input signal $x[n]$, even if the subband signals are not quantized.

We denote the Fourier transforms of the signals $x_l[m]$ and $x_h[m]$ resulting from the splitting by $X_l(e^{j\theta})$ and $X_h(e^{j\theta})$, where θ is the normalized frequency in radians. From the standard theory on sample rate conversion it follows that these Fourier transforms are equal to [54]

$$X_l(e^{j\theta}) = \tfrac{1}{2}[H_l(e^{j\theta/2})X_l(e^{j\theta/2}) + H_l(e^{j(\theta+\pi)/2})X_l(e^{j(\theta+\pi)/2})],$$

$$X_h(e^{j\theta}) = \tfrac{1}{2}[H_h(e^{j\theta/2})X_h(e^{j\theta/2}) + H_h(e^{j(\theta+\pi)/2})X_h(e^{j(\theta+\pi)/2})],$$

(13.1)

where $H_l(e^{j\theta})$ and $H_h(e^{j\theta})$ are the Fourier transforms of $h_l[n]$ and $h_h[n]$, respectively. The Fourier transform of the output signal $\hat{x}[n]$ is equal to

$$\hat{X}(e^{j\theta}) = G_l(e^{j\theta})X_l(e^{j2\theta}) + G_h(e^{j\theta})X_h(e^{j2\theta}),$$

(13.2)

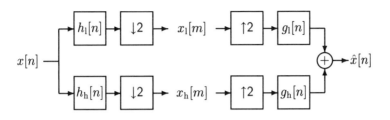

Figure 13.2 *Two-band subband filtering system.*

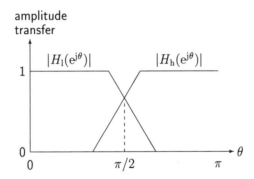

Figure 13.3 *Quadrature-mirror filters.*

where $G_l(e^{j\theta})$ and $G_h(e^{j\theta})$ are the Fourier transforms of the impulse responses $g_l[n]$ and $g_h[n]$. Substitution of the expressions for $X_l(e^{j\theta})$ and $X_h(e^{j\theta})$ gives

$$\hat{X}(e^{j\theta}) = \tfrac{1}{2}[G_l(e^{j\theta})H_l(e^{j\theta}) + G_h(e^{j\theta})H_h(e^{j\theta})]X(e^{j\theta}) +$$

$$\tfrac{1}{2}[G_l(e^{j\theta})H_l(e^{j(\theta+\pi)}) + G_h(e^{j\theta})H_h(e^{j(\theta+\pi)})]X(e^{j(\theta+\pi)}). \quad (13.3)$$

This equation can be written as

$$\hat{X}(e^{j\theta}) = T(e^{j\theta})X(e^{j\theta}) + A(e^{j\theta})X(e^{j(\theta+\pi)}), \quad (13.4)$$

where

$$T(e^{j\theta}) = \frac{1}{2}[G_l(e^{j\theta})H_l(e^{j\theta}) + G_h(e^{j\theta})H_h(e^{j\theta})] \qquad (13.5)$$

and where

$$A(e^{j\theta}) = \frac{1}{2}[G_l(e^{j\theta})H_l(e^{j(\theta+\pi)}) + G_h(e^{j\theta})H_h(e^{j(\theta+\pi)})]. \qquad (13.6)$$

$T(e^{j\theta})$ is the desired transfer function from $x[n]$ to $\hat{x}[n]$, and the term $A(e^{j\theta})X(e^{j(\theta+\pi)})$ is an alias component.

From the above expressions it can be seen that the output signal $\hat{x}[n]$ may be different from the input signal $x[n]$. Usually, it is required that, except for a delay, $x[n]$ and $\hat{x}[n]$ are equal. This implies that $T(e^{j\theta})$ should be equal to one, or to a pure delay, and that $A(e^{j\theta})$ should be identical to zero.

A well-known type of filter that satisfies these constraints very closely is the so-called quadrature-mirror filter (QMF)[104]. In a QMF splitting and reconstruction unit, the filters $h_l[n]$, $h_h[n]$, $g_l[n]$, and $g_h[n]$ are all derived from a common finite impulse response (FIR), symmetric, even-length lowpass filter $h[n]$, with N elements in the range $0 \le n \le N-1$, according to the following relations

$$\begin{aligned} h_l[n] &= h[n], \\ h_h[n] &= (-1)^n h[n], \\ g_l[n] &= 2h[n], \\ g_h[n] &= -2(-1)^n h[n]. \end{aligned} \qquad (13.7)$$

These relations imply that the transfer function of the highpass filters $h_h[n]$ and $g_h[n]$ are mirrored versions of those of the lowpass filters $h_l[n]$ and $g_l[n]$. This is schematically illustrated in Figure 13.3 for $h_l[n]$ and $h_h[n]$. It can be shown that with this choice of filters the aliasing term $A(e^{j\theta})$ is equal to zero [54], whereas the desired transfer is equal to

$$T(e^{j\theta}) = [|H(e^{j\theta})|^2 + |H(e^{j(\theta+\pi)})|^2]e^{-j\theta(N-1)}, \qquad (13.8)$$

where $H(e^{j\theta})$ is the Fourier transform of $h[n]$. Except for a delay of $N-1$ samples, the output signal $\hat{x}[n]$ can be made equal to the input signal $x[n]$ by choosing the impulse response $h[n]$ in such a way that

$$|H(e^{j\theta})|^2 + |H(e^{j(\theta+\pi)})|^2 = 1 \qquad (13.9)$$

for all θ. In most practical situations, we also want $h[n]$ as much as possible to approximate the ideal lowpass filter, i.e.

$$\begin{aligned} |H(e^{j\theta})| &\approx 1, \quad \text{for } 0 \le \theta \le \pi/2, \\ |H(e^{j\theta})| &\approx 0, \quad \text{for } \pi/2 < \theta \le \pi/2. \end{aligned} \qquad (13.10)$$

For $N > 2$ it is difficult to satisfy both constraints stated in (13.9) and (13.10). However, by using computer optimization methods, $h[n]$ can be designed such that

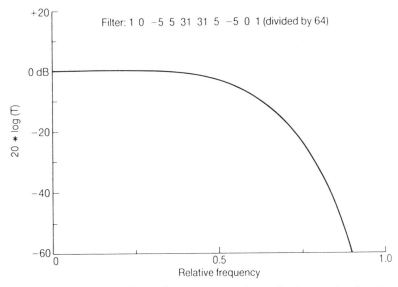

Figure 13.4 *Example of impulse response and amplitude transfer function.*

the difference between the input signal $x[n]$ and the output signal $\hat{x}[n]$ is very small. Figure 13.4 gives an example of the impulse response and the amplitude transfer function of a relatively simple filter $h[n]$ having only 10 elements [105]. The impulse response elements can all be represented with only 7 bits, which reduces the complexity of implementation of the filter. Figure 13.5 gives the magnitude of the overall transfer function of the QMF splitting and reconstruction unit using this filter.

By choosing filters other than QMFs it is in principle possible to design a subband filtering system that has perfect reconstruction. An example of such filters are the so-called conjugate quadrature-mirror filters (CQF) [73]. An important difference between QMF and CQF filters is that the former have linear phase characteristics, while the latter do not. In some applications, linear phase is required. An example, which will be discussed in Section 13.5, is the use for subband coding for compatible coding of high-definition television (HDTV) signals.

Up to now, only a simple two-band filter bank has been discussed. A division into more frequency bands can be obtained by repeatedly applying the two-band splitting and reconstruction sections of Figure 13.2. Figure 13.6 gives an example of a three-band filter bank, of which the middle- and high-frequency bands have equal bandwidth, and the lowest-frequency band has twice this width. Other methods for the realization of filter banks with more than two bands can be found in [55].

216   Subband coding of pictures

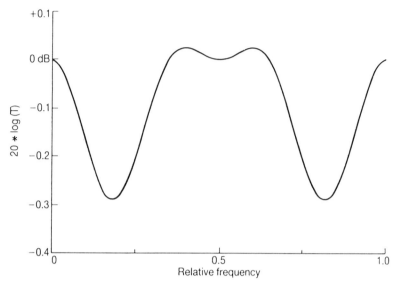

Figure 13.5 *Overall amplitude transfer function.*

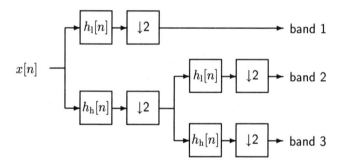

Figure 13.6 *Example of three-band subband filtering system.*

13.2.2   Filtering of pictures

Extension of one-dimensional subband filtering to pictures can be performed in a number of ways. The first and simplest method is to process pictures horizontally and vertically in a separable way. This method will be briefly treated in this section. The second, and more complex method is to use two-dimensional non-separable filters. This is now emerging in literature. Its treatment is beyond the scope of this book. Interested readers are referred to [106, 107, 108], for example.

The splitting of a picture into K horizontal spatial frequency bands and L vertical spatial frequency bands can be easily performed as follows. Each line of the

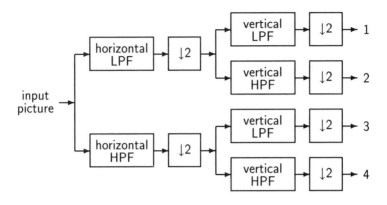

Figure 13.7 *Example of subband filtering of pictures: a 2×2 band system.*

picture is first split into K horizontal frequency bands. Subsequently, each column of the K resulting bands is split into L vertical frequency bands. Figure 13.7 shows the splitting configuration for $K = L = 2$, and Figure 13.8 shows how it divides the spatial frequency spectrum of the picture into four equal parts. Figure 13.9 shows the result of this splitting applied to the picture in Figure 9.2. Finally, Figure 13.10 gives examples of other possible splitting configurations [109, 110].

13.3 Quantization and coding

13.3.1 The effect of quantization

The effect of quantization of subband signals will be analysed in this subsection for the simple two-band splitting scheme of Figure 13.2. In the previous section it has been explained that, by using QMF, the output signal $\hat{x}[n]$ can be made a close replica of the input signal $x[n]$. This is only true if the signals $x_l[m]$ and $x_h[m]$ resulting from the splitting of $x[n]$ are transmitted to the decoder without any changes. In subband coding systems, however, the signals $x_l[m]$ and $x_h[m]$ are quantized before they are transmitted. How does the quantization influence the signal $\hat{x}[n]$?

Let us first assume that only the lowpass signal $x_l[m]$ is quantized to $\hat{x}_l[m]$, and

218 Subband coding of pictures

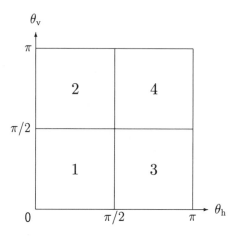

Figure 13.8 *Division of spatial spectrum using a 2×2 subband filter (θ_h=horizontal frequency, θ_v=vertical frequency).*

Figure 13.9 *Photograph of split picture.*

that the quantization introduces an additive quantization error $\varepsilon_1[m]$, i.e.

$$\hat{x}_1[m] = x_1[m] + \varepsilon_1[m]. \tag{13.11}$$

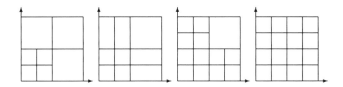

Figure 13.10 *Examples of other splitting topologies.*

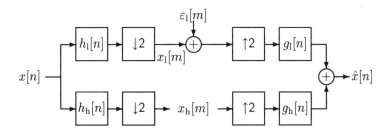

Figure 13.11 *Quantization of lowpass signal $x_1[m]$.*

This is illustrated in Figure 13.11. To understand the effect of quantization on the reconstructed signal $\hat{x}[n]$, we can study it in both the frequency domain and the discrete time domain. Let us start with the frequency domain.

Assuming that the quantization error signal $\varepsilon_1[m]$ has a Fourier transform $E_1(e^{j\theta})$, it can be easily derived that the Fourier transform of the output signal $\hat{x}[n]$ of Figure 13.11 is equal to

$$\hat{X}(e^{j\theta}) = [|H(e^{j\theta})|^2 + |H(e^{j(\theta+\pi)})|^2]e^{-j\theta(N-1)}X(e^{j\theta}) + 2H(e^{j\theta})E_1(e^{j2\theta}). \qquad (13.12)$$

If $h[n]$ is chosen properly we know that

$$|H(e^{j\theta})|^2 + |H(e^{j(\theta+\pi)})|^2 \approx 1, \qquad (13.13)$$

from which it follows that, approximately,

$$\hat{X}(e^{j\theta}) \approx e^{-j\theta(N-1)}X(e^{j\theta}) + 2H(e^{j\theta})E_1(e^{j2\theta}). \qquad (13.14)$$

From (13.14) it follows that if $h[n]$ is an ideal lowpass filter with cutoff frequency equal to $\pi/2$, the quantization of $x_1[m]$ introduces noise only in the lowpass band of $\hat{x}[n]$, i.e. for $0 \leq \theta \leq \pi/2$. For example, let us assume that uniform quantization with a step size equal to Δ_l is used to quantize $x_l[m]$. In that case the quantization noise power of $\varepsilon_l[m]$ is approximately flat with power equal to $\Delta_l^2/12$.

Very often, the performance of source-coding schemes is evaluated by the ratio between the power of the original, uncoded signal $x_l[m]$ and the power of the quantization noise in the coded signal $\hat{x}_l[m]$. If σ_l^2 is the power of the lowpass band $x_l[m]$, the signal-to-quantization-noise ratio in this band is equal to $12\sigma_l^2/\Delta_l^2$.

Similarly, it can be shown that the quantization of $x_h[m]$ introduces quantization noise only in the highpass band of $\hat{x}[n]$. This only holds if $h[n]$ is an ideal lowpass filter. If uniform quantization is used with a step size of Δ_h, the signal-to-quantization-noise ratio in this band is equal to $12\sigma_h^2/\Delta_h^2$, where σ_h^2 is the power in the highpass band $x_h[m]$.

A comparison of the signal-to-noise ratios in the lowpass and the highpass bands shows one of the main advantages of subband coding. By choosing different step sizes for each of the subbands, the power spectral density of the quantization noise can be varied from band to band. In that way, it can be optimally adapted to the properties of the receiver. The human visual system, for example, can tolerate significantly more quantization noise in the high spatial frequencies than in the low and middle ones. By quantizing the higher spatial frequencies more coarsely than the lower ones, a substantial reduction of the bit rate can be achieved. This is illustrated by the following example.

Let us assume that the picture of Figure 9.2 is split as indicated in Figure 13.8, and each of the subbands is uniformly quantized with a step size equal to 2. The first order entropy of the quantized signals, averaged over the four subbands, is equal to 3.42 bit per sample. However, the same perceptual quality can be obtained by quantizing bands 2 and 3 with a step size of, for example, 4, and band 4 with a step size of 8. By using these larger step sizes the first-order entropy reduces to 2.63 bit per sample.

For the analysis of the effect of quantization, it has been assumed that the lowpass filter $h[n]$ is ideal. In reality this will, of course, never be the case. Quantization of the lowpass signal $x_l[m]$ will then lead not only to quantization errors in the lowpass band, but also to errors in the highpass band. Similarly, the quantization of the highpass signals $x_h[m]$ will also introduce errors in the lowpass band.

In order to understand the local effect of a quantization error in $x_l[m]$ on the output signal $\hat{x}[n]$ it is more convenient to analyse the quantization error in the discrete time domain. Let us, for example, assume that the quantization error signal $\varepsilon_l[m]$ is equal to δ for $m = 0$, and is equal to zero for $m \neq 0$. Assume that without quantization of $x_l[m]$ the output signal of the subband system is approximately equal to $x[n]$. Then with quantization of $x_l[m]$ the output signal $\hat{x}[n]$ is approximately

$$\hat{x}[n] \approx x[n] + 2\delta h[n]. \tag{13.15}$$

From this expression we can see that $\hat{x}[n] \approx x[n]$ for $n < 0$ and for $n \geq N$. So the quantization error at time $n = 0$ only influences the output samples between time

$n = 0$ and $n = N - 1$, where N is the length of the filter $h[n]$.

For subband coding of pictures the following factors have to be taken into account. In pictures, a large spatial expansion of a quantization error made in a subband is undesirable, since this will generally increase the chance of detecting the quantization error. From (13.15) it follows that one must try to choose the filter length as small as possible. Furthermore, one should try to find an impulse response $h[n]$ that gives the least annoying effects in the spatial domain. It is known, for example, that the human visual system is very sensitive to so-called ringing near edges (echos of the edge). This ringing can be caused by a coarse quantization of the subbands in combination with a large amount of overshoot in the step response of the lowpass filter $h[n]$. Filters with a small amount of overshoot will therefore give the least annoying effects.

This section has only indicated the effect of the quantization of the lowpass subband signal $x_l[m]$. A similar treatment could be used for the highpass signal $x_h[m]$.

13.3.2 Bit allocation

In Subsection 12.3.3 the question of determining how accurately transform coefficients should be quantized has been dealt with. A similar treatment will be described in this subsection for quantization of subband signals.

Assume that the picture is split into K horizontal and L vertical spatial frequency bands, and that each subband signal is indicated as $s_{i,j}[m,n]$, where i,j with $i = 1, \ldots, K$ and $j = 1, \ldots, L$ indicates the number of the subband and $[m, n]$ the position of the subband samples. Assume further that the statistics of the picture are constant, and that the variance of subband signal $s_{i,j}[m, n]$ is equal to σ_{ij}^2.

The quantization of subband $s_{i,j}[m, n]$ will introduce a certain amount of quantization noise. Let $\sigma_{q,ij}^2$ be the variance of this noise. For simplicity, assume first that the effect of quantization of a subband on the perceptual picture quality is independent of the number of the band. In that case, the total amount of quantization noise

$$\sigma_q^2 = \sum_{i=1}^{K} \sum_{j=1}^{L} \sigma_{q,ij}^2 \tag{13.16}$$

is a good measure for the picture quality. Assume that the picture has to be coded at an average bit rate of R bit per sample. How accurately should each subband signal be quantized so that the total amount of quantization noise is minimized? Assume, for example, that Max-Lloyd quantization with fixed-length coding is used, with L_{ij} quantization levels for subband i, j. In [4] it is explained that the optimal choice for L_{ij} follows from

$$L_{ij} = 2^{R_{ij}}, \tag{13.17}$$

where

$$R_{ij} = R + \frac{1}{2}\log_2\left(\frac{\sigma_{ij}^2}{\left[\prod_{k=1}^{K}\prod_{l=1}^{L}\sigma_{kl}^2\right]^{1/KL}}\right). \qquad (13.18)$$

Expression 13.18 is similar to (12.25) from Chapter 12.

In reality the quantization of different subbands has different effects on the perceptual picture quality. It is known that much more quantization noise can be tolerated in high-frequency subband signals than in low-frequency ones. This can be incorporated in (13.18) by replacing σ_{ij} with $w[i,j] \cdot \sigma_{ij}$, where $w[i,j]$ is a weighting factor between 0 and 1 that expresses the sensitivity of the human visual system to quantization noise in subband $s_{i,j}[m,n]$, and by replacing σ_{kl} by $w[k,l] \cdot \sigma_{kl}$. The more sensitive the human visual system, the larger should be the value of $w[i,j]$.

As has been already explained in Subsection 12.3.3, pictures generally do not have constant statistics. This means that for an efficient representation of the subband signals, the bit allocation of (13.18) has to be made adaptive to the local statistics in the picture. This is generally referred to as adaptive bit allocation. A possible method is as follows. The subband signals can be divided into small, non-overlapping rectangular blocks of, for example, 4×4 samples. In each block the signal variance can be estimated. The bit allocation according to (13.18) can now be locally performed on blocks with similar positions in different subbands. Since the decoder must know the bit allocation, the estimated variances have to be transmitted to the decoder as side information. This side information will, of course, require a certain amount of bits. On the basis of the transmitted variances, the decoder can calculate the same bit allocation as the encoder.

13.4 Subband and transform coding

The reduction factors that can be obtained with subband coding of, for example, CCIR 601 TV pictures are in the range of 4–8, i.e. bit rates of about 0.5 to 2 bit/sample have been achieved. Of course, the bit rate depends on the difficulty of the picture, and on the splitting configuration, the quantization and the coding. Generally, highly detailed pictures require a higher bit rate than less detailed ones. Furthermore, the more sophisticated the quantization and the coding in the subband coding system, the higher the reduction factor that is obtained. Examples of subband coding systems can be found in [103, 109, 110, 111].

The reduction factors of subband-coding and transform-coding systems are similar. The coding artefacts, however, can be quite different, especially at low bit rates. Since transform coding operates on small blocks, the coding error is always confined to the area covered by one block. Especially at very low bit rates, this may lead to so-called blocking artefacts, i.e. separate blocks become visible. In subband-coding systems, coding at very low bit rates will generally lead to so-called ringing artefacts near sharp luminance transitions. These are most obvious for filters having a large amount of overshoot in the step response. Generally speaking, the longer the filters, the larger the area in which these effects are visible. It depends on the application as to which of these effects, blocking or ringing, is less annoying.

13.5 Specific applications

The interest in subband coding has recently increased, since it may play a key role in so-called compatible coding systems. The problem of compatibility is as follows. Currently, coding systems are designed for storage and transmission of digital TV signals (TV). However, within the next few years high-definition TV signals (HDTV) will probably be introduced. These signals will have twice as many lines as the digital TV signal, and twice as many samples, or more, per line. The question is how to design coding systems for these HDTV signals in such a way that coded HDTV can be decoded by TV coding systems.

Compatibility between TV and HDTV coding systems can be obtained by splitting the uncoded HDTV signal into a 'compatible' TV signal and one or more surplus signals. The surplus signals must contain the information needed to upgrade the TV signal to an HDTV signal. The compatible TV signal and the surplus signals are coded and transmitted separately. If the receiver only has a TV decoder, only the coded TV signal will be used. HDTV decoders will decode both the TV signal and the surplus signals, and combine them into an HDTV signal.

The splitting of HDTV can be performed with subband coding. For example, assume that the TV signal has exactly half as many lines and half as many samples per line as the HDTV signal. Then the splitting scheme of Figure 13.7 could be used to split each HDTV picture into four subbands, of which one is a 'compatible' picture with a resolution equal to that of TV pictures. This compatible picture can then be coded with already existing TV coding systems, while the other subbands can be coded using other techniques. In this way, two bit streams can be created, one containing the compatible pictures, and one containing the other three subbands.

In order to obtain a high-quality compatible TV picture, the lowpass filter in the subband-coding system has to have a linear phase characteristic. QMF is therefore especially suited to splitting. An example of a compatible HDTV coding system based on QMF can be found in [105]. Other interesting examples can be found in [112, 113, 114].

14
Pyramidal coding of pictures

14.1 Principle

This chapter is about pyramidal coding of pictures [115]. Like subband coding, pyramidal coding splits the input signal into a number of frequency bands one of which is a low-resolution version of the input signal. The frequency bands are quantized and coded. As was done for subband coding, pyramidal coding will first be explained for one-dimensional discrete-time signals. Then the explanation will be extended to cover the coding of pictures.

Figure 14.1 depicts a simple form of a pyramidal-coding system. In the encoder, the input signal $x[n]$ is lowpass-filtered with an impulse response $h_1[n]$ and subsequently subsampled by a factor of two. The resulting signal $x_1[m]$ is a low-resolution version of the input signal $x[n]$, having half its sample rate. From the standard theory on sample-rate conversion we know that the Fourier transform of $x_1[m]$ is equal to [54]

$$X_1(e^{j\theta}) = \frac{1}{2}[H_1(e^{j\theta/2})X(e^{j\theta/2}) + H_1(e^{j(\theta+\pi)/2})X(e^{j(\theta+\pi)/2})] \qquad (14.1)$$

The signal $x_1[m]$ is quantized, coded and transmitted to the decoder.

The next step in the encoder is the creation of a highpass signal $x_h[n]$. This is performed as follows. First, the lowpass signal $x_1[m]$ is upsampled by a factor of two and subsequently interpolated to $x_1'[n]$. The signal $x_1'[n]$ is subtracted from $x[n]$, which results in a highpass signal $x_h[n]$. The interpolation is performed with a lowpass filter with impulse response $h_2[n]$. The Fourier transform of $x_1'[n]$ is equal to

$$X_1'(e^{j\theta}) = H_2(e^{j\theta})X_1(e^{j2\theta}), \qquad (14.2)$$

where $H_2(e^{j\theta})$ is the Fourier transform of $h_2[n]$. Usually, $h_1[n]$ is a lowpass filter with a cutoff frequency equal to $\pi/2$, and $h_2[n]$ is chosen equal to $2h_1[n]$. In the remainder of this chapter, it is assumed that $h_1[n] = h[n]$, and $h_2[n] = 2h[n]$. The Fourier transform of $h[n]$ is indicated as $H(e^{j\theta})$.

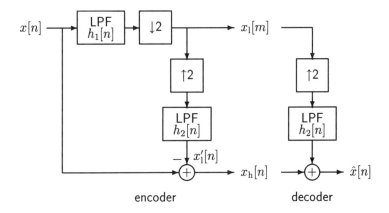

Figure 14.1 *Basic pyramidal filtering scheme.*

The Fourier transform of the highpass signal $x_h[n]$ is equal to

$$X_h(e^{j\theta}) = (1 - H^2(e^{j\theta}))X(e^{j\theta}) - H(e^{j\theta})H(e^{j(\theta+\pi)})X(e^{j(\theta+\pi)}). \tag{14.3}$$

This expression can be written as

$$X_h(e^{j\theta}) = H_h(e^{j\theta})X(e^{j\theta}) - A(e^{j\theta})X(e^{j(\theta+\pi)}), \tag{14.4}$$

where

$$H_h(e^{j\theta}) = 1 - H^2(e^{j\theta}) \tag{14.5}$$

is the Fourier transform of a filter with a highpass characteristic, and where

$$A(e^{j\theta}) = H(e^{j\theta})H(e^{j(\theta+\pi)}). \tag{14.6}$$

$A(e^{j\theta})X(e^{j(\theta+\pi)})$ is an aliasing term which is only identical to zero if the filter $h[n]$ is ideal. The highpass signal $x_h[n]$ is quantized and coded, and subsequently transmitted to the decoder.

In the decoder, the reconstructed signal $\hat{x}_1[m]$ is treated similarly to $x_1[m]$ in the encoder, i.e. the signal is upsampled and interpolated to a signal $\hat{x}'_1[n]$. This signal is then added to the reconstructed signal $\hat{x}_h[n]$ to obtain a reconstruction $\hat{x}[n]$ of the input signal $x[n]$. It is easy to show that, if no quantization is applied to $x_l[m]$ and $x_h[n]$, $\hat{x}[n]$ is equal to $x[n]$.

The basic scheme of Figure 14.1 can be repeated a number of times to split the input signal into a lowpass signal and a number of highpass signals. This is indicated schematically in Figure 14.2. Starting from a signal with a sample rate of M samples/second, the first application would result in a lowpass signal having

226 Pyramidal coding of pictures

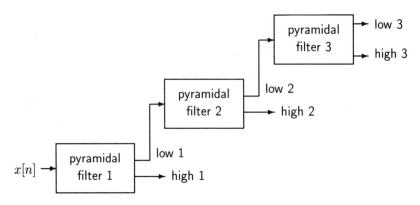

Figure 14.2 *Creation of a 'pyramid' of resolutions.*

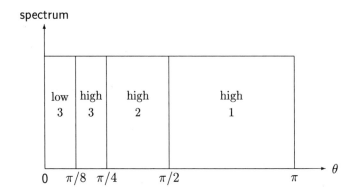

Figure 14.3 *Division of the signal spectrum by repeated splitting.*

$M/2$ samples/second, and a highpass signal with M samples/second. The lowpass signal can again be split into a lowpass signal, with $M/4$ samples/second, and a highpass signal with $M/2$ samples/second, and so on. Applying the basic scheme of Figure 14.1 three times would result in one lowpass signal and three highpass signals. Figure 14.3 shows how, in this case, the frequency spectrum is divided into four parts. In this figure it is assumed that ideal lowpass filters are used.

A generalization to coding of pictures is easily made by replacing the one-dimensional lowpass filter $h[n]$ by a two-dimensional lowpass filter $h[m, n]$, where m is the line index and n the column index. Assuming that this filter has a cutoff frequency $\theta_c \leq \pi/2$ both horizontally and vertically, a rectangular subsampling and upsampling by a factor of two both horizontally and vertically can be performed. In principle, however, any lowpass transfer function, with a corresponding subsampling raster, can be chosen for the two-dimensional lowpass filters.

It may be asked why the method described above is called pyramidal coding. This can be explained as follows. Figure 14.2 showed how a signal is repeatedly split. Suppose this procedure were repeatedly applied on a picture of M lines with N samples per line. The first splitting results in a lowpass picture of $M/2 \times N/2$, the second in a lowpass picture of $M/4 \times N/4$, and so on. If all lowpass pictures were placed on top of each other, the result would be something resembling a pyramid. Figure 14.4 shows the result of splitting the picture of Figure 9.2 once. Figure 14.4(a) shows the lowpass picture, and Figure 14.4(b) shows the highpass picture.

14.2 Quantization

In general, the signals $x_l[m]$ and $x_h[n]$ are first quantized and then transmitted to the decoder. The effect of quantization of these signals on $\hat{x}[n]$ will be described in this section. The extension of this description to pictures is left to the reader.

Let us first look at the quantization of $x_h[n]$ to $\hat{x}_h[n]$. We assume that quantization introduces an additive quantization error $\varepsilon_h[n]$, i.e.

$$\hat{x}_h[n] = x_h[n] + \varepsilon_h[n]. \tag{14.7}$$

Assuming $x_l[m]$ is not quantized, it follows that for the reconstructed signal

$$\hat{x}[n] = x[n] + \varepsilon_h[n], \tag{14.8}$$

or equivalently

$$\hat{X}(e^{j\theta}) = X(e^{j\theta}) + E_h(e^{j\theta}) \tag{14.9}$$

where $E_h(e^{j\theta})$ is the Fourier transform of $\varepsilon_h[n]$. Similarly, we assume a quantization error $\varepsilon_l[m]$ due to quantization of $x_l[m]$, i.e.

$$\hat{x}_l[m] = x_l[m] + \varepsilon_l[m]. \tag{14.10}$$

Then the Fourier transform of the output signal $\hat{x}[n]$ is equal to

$$\hat{X}(e^{j\theta}) = X(e^{j\theta}) + E_h(e^{j\theta}) + 2H(e^{j\theta})E_l(e^{j2\theta}), \tag{14.11}$$

where $E_l(e^{j\theta})$ is the Fourier transform of $\varepsilon_l[n]$.

Let us, for example, look at the simple case that $x_l[m]$ and $x_h[n]$ are both uniformly quantized with step sizes Δ_l and Δ_h, respectively. Assuming that these step sizes are sufficiently small and that $h[n]$ approximates an ideal lowpass filter, this uniform quantization leads to a total quantization noise power equal to $(\Delta_l^2 + \Delta_h^2)/12$ in the lowpass band of $\hat{x}[n]$ ($0 \leq \theta < \theta_c$), and equal to $\Delta_h^2/12$ in the highpass

228 Pyramidal coding of pictures

(a)

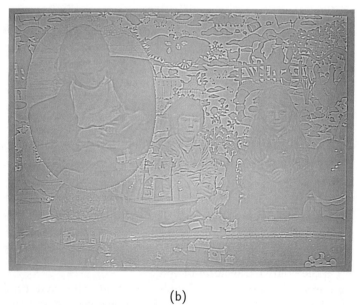

(b)

Figure 14.4 *(a) Lowpass picture, and (b) highpass picture.*

band ($\theta_c \leq \theta < \pi$). This power spectral density is illustrated in Figure 14.5 for

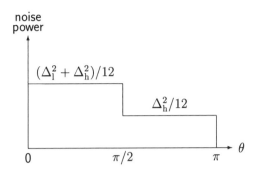

Figure 14.5 *Power spectral density of the quantization noise.*

$\theta_c = \pi/2$.

Let us now look at the effect of the quantization in the sample domain. The analysis of this effect is simple. Let us, as has been done for subband coding, assume that $\varepsilon_l[m] = \delta$ for $m = 0$ and $\varepsilon_l[m] = 0$ for $m \neq 0$. Then it follows for the reconstructed signal $\hat{x}[n]$ that

$$\hat{x}[n] = x[n] + \varepsilon_h[n] + 2\delta h[n]. \tag{14.12}$$

From this expression it can be concluded that the quantization error made in $x_l[m]$ is spread over an area which is equal to the length of the impulse response $h[n]$, whereas the quantization error made in $x_h[n]$ is not spread out. As in the case of subband coding, for picture source coding the filter $h[n]$ should be as short as possible, and it should not have a large amount of overshoot or undershoot. In case of a coarse quantization of the signal $x_l[m]$, filters with a large amount of overshoot will lead to clearly visible 'ringing' effects in the reconstructed picture.

Looking at the power spectral density given in Figure 14.5 we see that a simple uniform quantization of $x_l[m]$ and $x_h[n]$ leads to more quantization noise in the lowpass band than in the highpass band. In picture source coding, the opposite is usually required, since the human visual system can tolerate more noise in the higher frequency range.

A quantization noise spectrum that is better adapted to the characteristics of the human visual system can be obtained in at least two ways. First, it is possible to encode the signals $x_l[m]$ and $x_h[n]$ with source-coding methods that are better capable of shaping the quantization noise spectrum than uniform quantization. For example, the signals could both be coded with transform coding. Second, it is possible to feed the lowpass quantization error $\varepsilon_l[m]$ into the highpass signal $x_h[n]$, so that this error can be compensated for. This is indicated in Figure 14.6. Let us assume that quantization of the highpass signal $x_h[n]$ in this figure leads to a quantization error $\varepsilon'_h[m]$ with Fourier transform $E'_h(e^{j\theta})$. It can be easily shown

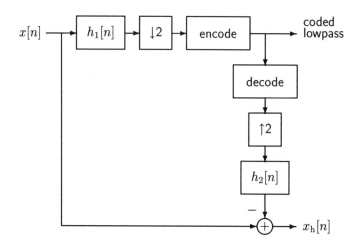

Figure 14.6 *Feedback of lowpass quantization errors into the highpass signal.*

that the Fourier transform of the output signal $\hat{x}[n]$ is then equal to

$$\hat{X}(e^{j\theta}) = X(e^{j\theta}) + E'_h(e^{j\theta}) \tag{14.13}$$

The frequency characteristics of the coding error in $\hat{x}[n]$ can now be controlled by quantization of the highpass signal $x_h[n]$.

14.3 Pyramidal and subband coding

Pyramidal coding and subband coding have much in common. They both split a picture into a lowpass picture, and one or more highpass pictures. There are, however, also a number of important differences between the two techniques.

The first difference is that the filter bank in a subband-coding system does not increase the number of samples to be coded. If the input picture has a size of $M \times N$, the number of samples that have to be quantized and coded is equal to MN. In a pyramidal coding system, more samples have to be quantized and coded. For example, application of the basic system of Figure 14.1 both horizontally and vertically results in a lowpass picture of size $M/2$ by $N/2$, and a highpass picture of M by N. This means that $5MN/4$ samples have to be processed. In order to obtain the same bit rate as in subband coding, pyramidal coding may therefore need somewhat more complex coding techniques.

The second difference is that subband coding can more easily divide the spectrum of the picture into small frequency bands. As a result, it is more easy to vary the quantization accuracy in different regions in the spectrum. Noise shaping in a pyramidal-coding system generally requires more complex quantization techniques,

The third difference is that in a pyramidal-coding system there is more freedom to determine the shape of the lowpass filter. For example, short filters with a small amount of overshoot can be chosen. This is especially of advantage in applications where it is required to control carefully the characteristics of the low-resolution picture. One of these applications, compatible HDTV coding, will be described in the next section.

14.4 Specific applications

Pyramidal coding is of specific interest for those applications in which it is required to have a hierarchy of pictures at different resolutions. One example is the storage of medical pictures, such as x-ray pictures, in a medical data base. On retrieval of those pictures, the physician often wants to search rapidly through a large set of pictures. A low-resolution version of the picture may then be sufficient for its recognition. The physician could therefore confine attention to these low-resolution versions, which can be decoded faster than the high-resolution pictures, until the appropriate picture is found. Then, this picture can be decoded to its full resolution.

Pyramidal coding can also be used in so-called compatible coding systems. The problem of compatibility has already been described in Section 13.5 and an example of a compatible HDTV coding system has been given there. A compatible HDTV coding system can also be realized using pyramidal coding. For example, assume that the HDTV signal has to be split into a compatible standard-definition TV signal and one surplus signal, and that the TV signal must have exactly half as many lines and half as many samples per line as the HDTV signal. The splitting scheme of Figure 14.1 can be used both horizontally and vertically. It splits each HDTV picture into a compatible TV picture and one surplus picture. The compatible picture can be coded with already existing TV coding systems, while the surplus picture can be coded using other techniques.

15
Source coding of picture sequences

15.1 Introduction

A number of picture coding techniques have been described in Chapters 10–14 of Part II. A feature common to these techniques is that they operate on one picture.

Sometimes a picture is part of a sequence of pictures. Examples are digital television signals. In Europe, for example, the CCIR 601 TV signal contains 25 pictures per second. Each picture consists of 576 lines with 720 luminance and 720 chrominance samples per line [81]. The pictures are usually referred to as frames. Of course, each of these frames can be individually coded using the techniques of Chapters 10–14. However, the coding efficiency can be significantly increased by taking advantage of the temporal redundancy in the sequence.

Figure 15.1 illustrates the presence of temporal redundancy. It shows the differences between two successive frames of the CCIR 601 TV signal of Figure 9.2. To make the differences clearly visible, they have been multiplied by a factor of four, and a value of 128 (middle grey luminance) has been added to them. Figure 15.2 shows the histogram of the differences. Both Figures 15.1 and 15.2 show that the differences are small. Furthermore, the histogram has a somewhat peaked form around zero, indicating that it is possible to represent the differences efficiently with relatively few bits, using e.g. variable-length coding techniques.

In order to benefit from temporal redundancy, two or more successive pictures have to be coded in combination. Since a picture in a TV sequence is often called a frame, techniques that code pictures in combination are generally called interframe-coding techniques. They are sometimes also referred to as spatio-temporal source-coding techniques. Techniques that encode each frame separately are usually referred to as spatial or intraframe-coding techniques.

Many different interframe-coding techniques have been proposed in literature. Interesting reviews can be found in, for example, [116, 117]. In the next few sections, one of the currently most popular techniques will be introduced step by step.

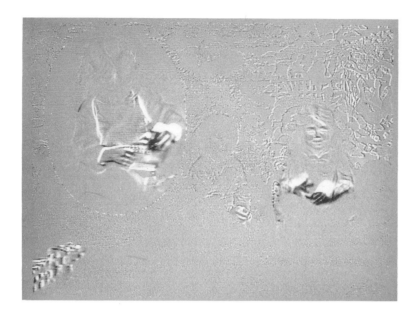

Figure 15.1 *Differences between two successive frames of a CCIR 601 TV signal.*

15.2 Temporal DPCM

In the remainder of this chapter the nth sample on the mth line of the ith picture in a sequence is indicated as $p_i[m,n]$. Let us consider the sample values as a function of the time index i, for certain fixed values of $m = m_c$ and $n = n_c$, and let us indicate these sample values as $x[i]$, i.e.

$$x[i] = p_i[m_c, n_c] \tag{15.1}$$

The one-dimensional source-coding techniques introduced earlier in this book can be applied to encode this signal $x[i]$, for every combination of values of m_c and n_c. This is usually referred to as temporal coding. For example, subband coding could be used. However, application of subband coding would require storage of a relatively large number of pictures. Furthermore, it would introduce a significant delay between encoder input and decoder output.

A technique that is less complex in terms of required storage capacity and in terms of introduced delay is temporal DPCM. Figure 15.3 shows a general block diagram of a temporal DPCM system. Note that this figure is very similar to that for spatial DPCM in Figure 11.1 in Chapter 11.

Briefly summarized, temporal DPCM operates as follows. For every sample value $p_i[m,n]$ a prediction $\tilde{p}_i[m,n]$ is made, based on the sample values in the previously reconstructed picture $i-1$. The samples $\hat{p}_{i-1}[m,n]$ of this previously

234 Source coding of picture sequences

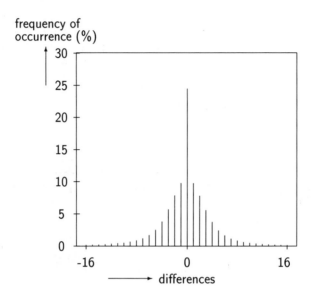

Figure 15.2 *Histogram of frame differences.*

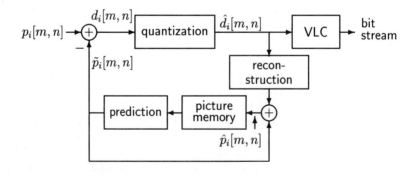

Figure 15.3 *Temporal DPCM: encoder.*

reconstructed picture are assumed to be stored in a memory. Now, instead of transmitting the samples $p_i[m,n]$ to the decoder, the differences

$$d_i[m,n] = p_i[m,n] - \tilde{p}_i[m,n]. \tag{15.2}$$

between these samples and their predictions are transmitted. The difference signal is often called the prediction error signal. Before transmission, this signal is quantized to $\hat{d}_i[m,n]$, and then coded to binary codewords. Variable-length coding is usually used.

In the decoder, the same prediction $\tilde{p}_i[m,n]$ is made, and is added to the reconstructed prediction error $\hat{d}_i[m,n]$. This yields a reconstructed signal

$$\hat{p}_i[m,n] = \tilde{p}_i[m,n] + \hat{d}_i[m,n]. \tag{15.3}$$

Since both the encoder and the decoder derive their predictions from the previously reconstructed picture, both have to contain a memory in which the previously reconstructed picture is stored.

The questions to be answered now are: how should the prediction be performed, and how should the prediction error signal be quantized and coded? These questions will be dealt with in the next few sections.

15.3 Prediction techniques

In spatial DPCM a predicted sample value is determined from a linear combination of a number of reconstructed samples, in a small area round the sample to be predicted. Figure 11.4 in Chapter 11 illustrates this method of prediction. A similar procedure can be used in temporal DPCM. In that case, the predicted sample value $\tilde{p}_i[m,n]$ is determined from a linear combination of reconstructed samples in picture $i-1$. For example, a very simple prediction could be to take $\tilde{p}_i[m,n] = \hat{p}_{i-1}[m,n]$, i.e. to take the reconstructed sample at the same position in the previous picture as a prediction. The efficiency of this type of predictions, however, is relatively low. Especially in regions containing temporal luminance changes, caused, for example, by moving objects, this type of prediction would result in large prediction errors, and thus in a relatively high bit rate.

A prediction technique that has proved to have a greater accuracy is motion-compensated prediction [117]. The principle of motion-compensated prediction is as follows. The picture is first divided into a large number of small regions. Division into non-overlapping blocks of, for example, 8×8 or 16×16 samples is very often used. In principle, however, any shape of any size is possible. Then the 'most similar' block in the previously reconstructed picture is determined for each of the blocks. Since the decoder must be able to determine exactly the same prediction, information about the relative position of this most similar block has to be transmitted to the decoder as side information. This information is often called the

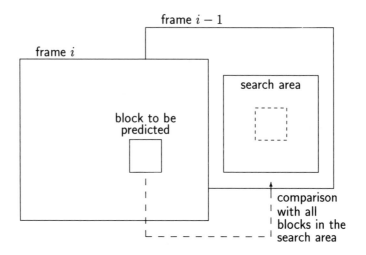

Figure 15.4 *Motion-compensated prediction: full-search block matching.*

motion vector, since it indicates how the information in the block to be coded has 'moved' from a certain position in picture $i-1$ to another position in picture i.

Figure 15.4 illustrates a possible technique to determine the 'most similar' block in the previously reconstructed picture. This technique is known as full-search block matching. The block to be coded is compared to all possible blocks inside a certain region of the previously reconstructed picture. This region is called the search range or the search area. In order to find the most similar block, some measure of similarity is required. Very often the mean-square error (MSE) is used, i.e. the block in the reconstructed picture that has the smallest MSE compared with the block to be coded is chosen as the most similar block, and is used as the prediction. Figure 15.5 shows the result of full-search block matching applied to the TV sequence of which Figure 9.2 shows the first frame. Figure 15.4 shows the displaced-frame differences between frame 2 and frame 1 of the TV sequence.

The determination of the most similar block is usually referred to as motion estimation. Full-search block matching is only one of the many existing motion-estimation techniques. Others can be found in for example [117, 118, 119, 120].

The prediction error, which is often called the displaced-frame difference (DFD), is quantized and coded. Figure 15.6 shows a block diagram of a temporal DPCM scheme using motion-compensated prediction.

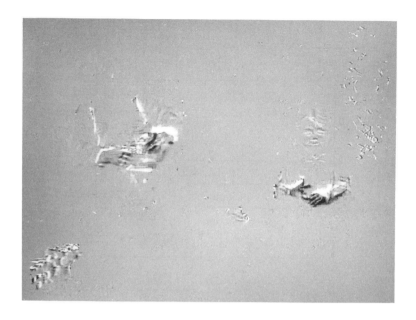

Figure 15.5 *Displaced-frame differences between two successive frames of a CCIR 601 TV signal.*

15.4 Quantization and coding

There are various techniques for quantization and coding of the prediction error signal. In principle, all the techniques treated earlier in this book can be applied. A method that has become popular in the last few years is depicted in Figure 15.7. In this figure, the prediction error signal is encoded with a transform coder such as that explained in Chapter 12 of Part II. For example, blocks of 8×8 samples of the prediction error signal can be first transformed to a spatial frequency domain using the DCT, and then weighted, quantized and coded. Since the source encoder in Figure 15.7 needs to have a reconstructed picture, it must also contain a DCT source decoder to derive this signal.

Especially for picture sequences which do not change rapidly with time, a scheme as given in Figure 15.7 can be very efficient. The reduction factor obtainable for such a signal may be more than twice as large as that obtainable with separate transform coding of each picture. For common TV signals, the compression ratio will be between one and two times larger.

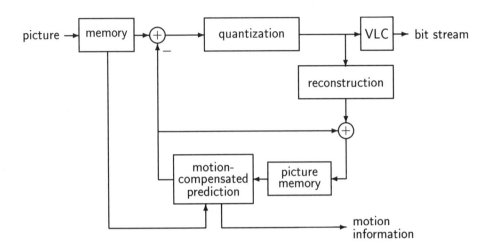

Figure 15.6 *Temporal DPCM with motion-compensated prediction.*

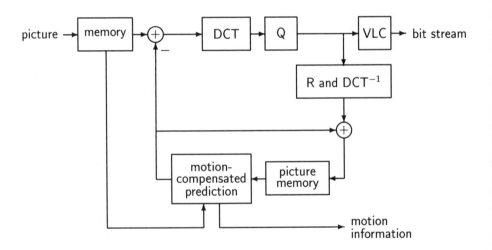

Figure 15.7 *Motion-compensated interframe DCT coding.*

15.5 Specific applications

Motion-compensated interframe codecs were first applied in so-called videophone and videoconferencing systems. They are currently under study for the recording and transmission of TV and HDTV signals. During the last few years the Motion Picture Experts Group (MPEG) of the International Standardization Organization (ISO) has been working on standards for interframe coding of video signals with bit rates between 1.2 and about 10 Mbit/s [121]. For CCIR 601 TV signals, for example, the reduction factors are then between about 16 and 140.

Part III

Conclusions

16
In conclusion to Part I

16.1 Introduction

Part I dealt with a theory behind source-coding or bit-rate-reduction systems. We have mainly restricted ourselves to source coding of 'physical' signals such as speech, music and picture sequences. Such a signal is input to the source encoder and output from the source decoder. Output from the source encoder and input to the source decoder are sequences of bits. Because we have restricted ourselves to waveform coding, the output of the source decoder is an approximation, e.g. quadratic, of the source decoder's input,

Firstly, it was argued that source coding is feasible for two reasons. The first is that signals are often represented with more bits than are necessary to fully describe them. The signal's representation is then said to be redundant. The second reason is that signals are often described with more detail than is required by the human observer at the destination. The superfluous details in the representation are called irrelevancy. The bit rate is reduced by removing irrelevancy and redundancy. This is the general approach to waveform coding of signals such as speech, music and picture sequences.

Source-coding systems are explained in terms of three distinct basic blocks that can be identified in each source-coding system. These blocks are coder, quantization and transformation blocks. They can be found in the source encoder and have counterparts in the source decoder which are called decoder, reconstruction and inverse transformation blocks, respectively. Coder, quantization and transformation blocks are introduced in Chapter 1. They are discussed in greater detail in Chapters 3 to 5.

For a good or optimal performance, the basic blocks must be tuned to signal properties. In most practical cases, these properties are not constant as a function of time. Therefore the basic blocks are often adaptive. Adaptivity in source coding is discussed in Chapter 6. A special case of adaptivity in source coding is bit-rate control. This is required when the combination of a transformation, quantization and coder block produces a bit stream with a variable bit rate and a fixed bit rate is required. Bit-rate control methods are discussed in some detail in a separate section of Chapter 6.

The main results of Chapter 3 to 5 are summarized in the following sections.

16.2 Redundancy, irrelevancy and distortion

It is not obvious that subsequent source encoding and decoding lead to signals of acceptable quality, but there are two reasons why source coding can be feasible. They are discussed below.

The first reason is that many signals are represented with more bits per second than is really necessary. In terms of information theory, such a representation is redundant. A reduction of the bit rate can be obtained by using another representation with fewer bits per second. In this case source coding is said to remove the redundancy. This is a reversible process: the source decoder can always recover the original input signal.

This type of source coding is sometimes called lossless source coding. An example of lossless coding discussed in Chapter 3 is called variable-length coding because it codes symbols with a fixed length with codewords of variable length. It is useful if the fixed-length symbols have unequal probabilities. Other lossless coding methods exist that can also be used to reduce the bit rate of a redundant source.

The second reason for the feasibility of source coding is that a perfect reconstruction of the input signal is not always required after source decoding, but that differences up to a certain degree between the input and the output are either acceptable or imperceptible. If this is the case, the bit rate can be reduced by finding a less accurate representation that requires fewer bits per second. The fact that source-encoder input and source-decoder output are different is, in this case, irrelevant to the observer. An encoder working in this way is said to remove irrelevancy from the signal's representation. This is irreversible. The amount of deviation between the input signal and the decoded signal is called the distortion.

16.3 Basic blocks of a source-coding system

Removal of irrelevancy from a signal's representation, which reduces the accuracy of the representation, is an operation that can be found in almost every source-coding system. It is generally called quantization. Quantization involves two types of quantization blocks: the quantizer, denoted by Q, in the source encoder; and the reconstructor, denoted by R, in the source decoder. The input of a quantizer is a real number or an integer that is very accurately represented with a large number of bits. The quantizer maps its input on to a finite set of so-called quantizer-output symbols. Because there are fewer distinct quantizer-output symbols than quantizer-input values, they can always be represented with a smaller number of bits. In the source decoder the reconstructor converts the received quantizer-output symbols into numbers in the same representation as the original quantizer-input values. These numbers are called quantization levels. Because there are fewer distinct quantization levels than distinct quantizer-input values, accuracy of representation has been lost.

The quantizer is essentially a source that produces discrete symbols. If the quantizer-output symbols have unequal probabilities, it is advantageous to code them with a variable-length code. To this end source-coding systems contain another pair of blocks: a coder block, schematically denoted by C, in the source encoder; and a decoder block, schematically denoted by D, in the source decoder. The coder reduces the bit rate by exploiting redundancy. The decoder performs the inverse mapping and produces the original fixed-length symbols. The minimum achievable bit rate is given by an entity called the entropy, which is a function of the probabilities of the quantizer-output symbols. A well-known code is the Huffman code.

In a simple coder-decoder pair, single quantizer-output symbols are mapped on to binary codewords. In such a system the coding complexity depends on the number of distinct quantizer-output symbols. In these simple systems possible statistical dependencies between quantizer-output symbols are not exploited. In order to exploit these dependencies, or in order to approach the entropy more closely, groups of quantizer-output symbols can be mapped on to binary codewords. This will yield a better removal of redundancy, but the coder's complexity will increase exponentially with the number of quantizer-output symbols grouped together.

With the blocks Q, C, D and R we can construct a source-coding system such as shown in Figure 5.1 on page 87. This source-coding system would not be suitable for speech, music or pictures, for the following reason. In order to keep the distortion in these signals below an acceptable level, the number of quantization-output symbols must be relatively high. For pictures it is in the range 2^6–2^8, for speech signals it is in the range 2^8–2^{12} and for music it is in the range 2^{13}–2^{16}. In order to achieve a low bit rate, dependencies between quantizer-output symbols must be exploited by coding groups of quantizer-output symbols. Because of the necessarily high number of quantizer-output symbols, this would result in an unacceptably high coding complexity.

The transformation block T changes the representation of the signal. In transform-coding systems for pictures or audio signals, for instance, a frequency-domain representation is obtained, which is then coded by the system shown in Figure 5.1 on page 87. The transformation output is often referred to as transformation coefficients. The coded representation is converted back into the original representation by the inverse transformation block. Well-known transformations are orthogonal transforms, subband filtering and predictive filters. Orthogonal transforms, such as the discrete cosine transform, and subband filtering are often used for picture and music coding. Predictive filtering is mainly used in speech coding. The advantages of using transformations are discussed in the following paragraphs.

They have been derived in Part I for Gaussian signals and uniform quantizers, but are usually confirmed by experiences with non-Gaussian signals and other types of quantizers. However, in such cases the results are mathematically less tractable.

The first advantage of using transformations is that it overcomes the unsuitability of the simple system shown in Figure 5.1 on page 87. The following explanation is adequate for transform coding and for subband coding, which are nowadays seen as closely related. The story for predictive filtering is somewhat different. The reader is referred to Chapter 5 for more details. The difference lies in the fact that the basis functions of transform and subband filtering are, at least by good approximation, orthogonal, whereas the basis functions of a predictive filtering operation are not.

We compare a source-coding system with a transformation, as shown in Figure 5.2 on page 88 with the simple source-coding system of Figure 5.1. The quantization is such that both systems produce the same distortion and approximately the same bit rate. Good transformations concentrate the signal power into a few transformation coefficients that are also, by good approximation, statistically independent. It is shown in Chapter 5 that the concentration of power into a small number of transformation coefficients has the effect of reducing the average number of quantizer-output symbols. This, combined with the fact that the quantized transformation coefficients can be coded independently, substantially reduces the coding complexity of the system with the transformation compared with the system without a transformation. Even if groups of quantized transformation coefficients are coded together, which will further reduce the bit rate by at most 1 bit per sample, the complexity will still be lower than that of the simple system.

It is important to notice that if a transformation has the concentration property, but is not optimal in the sense that the transformation coefficients are independent, a reduced bit rate at a low complexity can still be achieved.

The second advantage of using transformations is a perceptual one. Direct quantization of the input samples has, to a large extent, the effect of adding uncorrelated noise to the signal. Both the human eye and the human ear have a certain sensitivity to this type of noise, which determines the acceptable noise level. It has been found that, when the right transformation is used, e.g. a transformation to the frequency domain for audio signals, the human observer is far less sensitive to distortion of some of the transformation coefficients than to uncorrelated noise. This implies that for some of the transformation coefficients a much lower signal-to-noise ratio can be tolerated, which allows them to be coded with a further reduced number of bits. This procedure is called noise-shaping. In some cases the sensitivity of the human observer to errors in a particular transformation coefficient is so low that it can be completely discarded and does not require any bits at all.

Of particular interest are those transformations that on the one hand transform the signal into a representation that allows noise-shaping and on the other hand have the concentration property. We believe that they will remain a topic for research for quite some time.

17

In conclusion to Part II

Part II of this book has illustrated how the principles explained in Part I can be used in practice to reduce the bit rate of real-life digital signals such as digital audio, digital pictures and digital picture sequences. A variety of source-coding methods for these signals has been described. How the transformation, quantization and coding are performed has been explained, and the error introduced by the source coding has also been analysed.

Chapter 8 has discussed subband-coding for audio signals and has paid special attention to a method which reduces the bit rate of compact disc digital audio signals by a factor of four without introducing an audible loss of sound quality. In this method, the audio signal is first divided into 32 frequency bands of equal width. Subsequently, each of these bands is quantized in such a way that the quantization noise is below the threshold of audibility. A similar system is used in the recently introduced Digital Compact Cassette (DCC®) audio recorder [62]. It is part of a more extensive audio source-coding scheme that has been drafted as an ISO standard [63, 77]. This more extensive scheme is also being considered for future digital audio broadcasting.

Chapters 9–15 have described a number of source-coding methods for digital pictures. The adaptive pulse-code modulation (APCM) technique described in Chapter 10 and the predictive coding technique (DPCM) described in Chapter 11 are the techniques with the lowest complexity. They operate directly on the picture samples or on a low-complexity prediction of these samples. They give moderate reduction factors of about, for example, 2-4 for CCIR 601 digital TV pictures. Techniques such as these are now used in several commercially available picture-coding systems. For example, the Philips division TRT in France has recently introduced a television distribution codec based on lowpass prefiltering and subsampling of the colour component of the television signal and DPCM of the luminance component. The codec operates at 34 Mbit/s, audio included [122].

The transform-coding, subband-coding and pyramidal-coding techniques described in Chapters 12–14 each first transform the picture to some spatial frequency domain. The frequency components resulting from the transformation are then quantized and coded. These techniques are more complex than APCM and

DPCM. In return, they facilitate a substantially larger reduction of the bit rate. For example, transform coding can reduce the bit rate of CCIR 601 digital TV pictures by a factor of about 4–8. At reduction factors up to 4 almost no visible distortion is introduced. This makes transform coding at these reduction factors suitable for the more professional applications such as cost-effective digital magnetic recording of TV and HDTV in television studios. At the higher reduction factors a good consumer picture-quality is still obtained.

Transform coding using the discrete cosine transform (DCT) is currently the most popular technique for source coding of still pictures and for intraframe coding of picture sequences such as digital video. A still-picture DCT-based method has recently been standardized by the Joint Picture Experts Group (JPEG) of the International Standardization Organization (ISO). The first source-coding chips and small printed-circuit boards with a complete source coder, for example, for personal computers, are now coming on to the market. Furthermore, experimental digital magnetic video recording systems based on intraframe DCT coding have been reported in the literature [101].

Chapter 15 has described a technique for source coding of picture sequences. This technique takes advantage of the facts that successive pictures in a sequence generally contain a significant amount of similarity and that the relative displacement of objects can be predicted reasonably well. Therefore, so-called motion-compensated predictive coding methods are used temporally. The prediction error is source coded with a block-based transform coding as described in Chapter 12. The performance of this technique is significantly better than that of the picture-coding methods described in Chapters 10–14. Its complexity, however, is also much higher. The motion compensation requires the storage of two or more frames as well as a large number of calculations.

A method similar to that described in Chapter 15 is used in videophone and videoconferencing systems. The source-coding method for these systems has now been standardized and named H2.61. It allows real-time video transmission via telephone links at a bit rate of $p \times 64$ kbit/s ($p = 1, \ldots, 30$). Systems conforming to this standard are now commercially available.

A more sophisticated form of motion-compensated predictive picture-sequence coding has recently been proposed for standardization by the Motion Pictures Experts Group (MPEG) of the ISO. The method, which is referred to as MPEG1, can code medium-resolution video at bit rates up to 1.2 Mbit/s. A similar method for coding CCIR 601 resolution video at bit rates of 4–10 Mbit/s is currently under investigation within MPEG. The first MPEG1 encoder and decoder chips are now coming on to the market.

18
Future developments

The audio and picture-coding techniques described in Part II have reached a certain state of maturity. Many products incorporating these digital source-coding methods can therefore be expected in the coming 10 years. These products will probably be the first of a series of new storage, transmission and communication systems.

The need for digital storage and transmission of audiovisual data is expected to grow in the near future. Systems in which storage and transmission of audio, speech, video and data are integrated are anticipated. They are generally referred to as multi-media systems. Research into new, more powerful source-coding techniques is required to enable a major step forward in obtainable reduction factors. Only then can cost-effective storage and transmission be guaranteed. Within the academic world and the research laboratories of major electronic and telecommunication companies research into these new, more powerful techniques has already started.

We see three, basically distinct, approaches to achieve source coding at lower bit rates. They will be discussed in the following three paragraphs.

The first, and least spectacular approach, is the use of better signal transformations, which supply a better decomposition of the signal into a small number of, preferably, perceptually-relevant components. It requires a better use of signal processing techniques and a better knowledge of human perception. This approach may be described as an attempt to improve existing systems. It will yield the quickest success, but will also demonstrate the limitations of existing systems. Already emerging examples of better transformations are overlapped orthogonal transforms [123], wavelet decomposition [124] and signal-adaptive subband filtering [125]. The use of signal-adaptive transformations [80] is also relatively new.

One of the limitations of existing source-coding systems is that they are based on waveform coding and that at some point quantization noise is introduced. This quantization noise is shaped in such a way that it causes minimal disturbance. Preferably it should be kept inaudible or invisible. However, only a limited amount of noise can be hidden from the human observer. If an attempt is made to further decrease the bit rate, it will become either visible or audible. The second approach to source coding at lower bit rates, which can be used in certain low-quality applications, accepts that noise will become perceptible and tries to shape it as well as possible. This is already done in some speech coders, but the amount of audible noise that is introduced is still quite low. If the amount of noise increases, we have to find noise-shaping methods that work well above the threshold of perception. This implies that we need quality measures for audio and video signals that not only inform us whether noise is perceptible, but also evaluate the perceptual quality of noisy signals. This is largely an open research problem. This approach can also be described as an extension of existing methods, but it requires far more knowledge of human perception. Unfortunately, it will only bring a solution in the case of low-quality applications.

The third approach drops the concept of waveform coding either completely or to some extent. In this approach, called model-based source coding or parameter coding, the signal is first modelled as accurately as possible. The model parameters are source coded. For example, a picture may first be separated into different regions with different textures, and then only the shapes of the regions and the characteristics of the textures are transmitted. A crude, older version of such a model-based system is the old speech vocoder in which only information about an excitation signal, e.g. fundamental frequency or noise variance, and about a synthesis filter were coded and transmitted at regular time intervals. With these parameters, the decoder tried to regenerate the signal. Newer versions of such coders, which perform waveform-coding as well as model-based coding, can be found in [126]. Generally, this is an entirely open research area. We don't find it very likely that completely model-based coders will emerge very soon. It is more probable that source-coding systems will gradually become increasingly model-based.

In conclusion we believe that the lower bounds of achievable bit rates for audio and picture source coding have certainly not been reached and that many exciting developments can be expected in the coming 20 or so years. We hope to have given the reader an insight into the area of source coding that will allow him to follow or even contribute to the developments in this area.

References

[1] J.B.H. Peek. Communication aspects of the compact disc digital audio system. *IEEE Communications Magazine*, 23(2):7–15, 1985.

[2] K.A. Schouhamer Immink. *Coding Techniques for Digital Recorders*. Prentice Hall, London, 1991.

[3] J.C. Mallinson. *The Foundations of Magnetic Recording*. Academic Press, London, 1987.

[4] N.S. Jayant and P. Noll. *Digital Coding of Waveforms*. Prentice Hall, Englewood Cliffs, New Jersey, 1984.

[5] A.N. Netravali and B.G. Haskell. *Digital Pictures: Representation and Compression*. Plenum Press, London, 1988.

[6] T. Berger. *Rate Distortion Theory*. Prentice Hall, 1971.

[7] R.M. Gray. *Source Coding Theory*. Kluwer Academic Publishers, Boston, 1990.

[8] A.W.M. van den Enden and N.A.M. Verhoeckx. *Discrete-Time Signal Processing: An Introduction*. Prentice Hall, Englewood Cliffs, New Jersey, 1989.

[9] L.R. Rabiner and B. Gold. *Theory and Application of Digital Signal Processing*. Prentice Hall, Englewood Cliffs, New Jersey, 1975.

[10] A. Papoulis. *Probability, Random Variables, and Stochastic Processes*. McGraw-Hill Kogakusha, Ltd., Tokyo, 1965.

[11] M. Marcus and H. Minc. *Introduction to Linear Algebra*. MacMillan, New York, 1965.

[12] G.H. Golub and C.F. van Loan. *Matrix Computations*. North Oxford Academic Publishing, Oxford, England, 1983.

[13] M.B. Priestley. *Spectral Analysis and Time Series*. Academic Press, London, 1981.

[14] H.L. van Trees. *Detection, Estimation and Modulation Theory*. John Wiley and Sons, New York, 1968.

[15] A.B. Carlson. *Communication Systems*. McGraw-Hill, New York, 1988.

[16] F.J.A. MacWilliams and. N.J.A. Sloane. *The Theory of Error-Correcting Codes*. North-Holland, Amsterdam, 1977.

[17] G.C. Clark Jr. and J.B. Cain. *Error-Correction Coding for Digital Communications.* Plenium Press, New York, 1981.

[18] J.L. Flanagan. *Speech Analysis Synthesis and Perception.* Springer-Verlag, Berlin, 1972.

[19] A. Jacquin. Fractal image coding based on a theory of iterated contractive image transformations. In M. Kunt, editor, *SPIE Visual Communication and Image Processing '90*, SPIE Processing Series, pages 227–239. SPIE, Bellingham, 1990.

[20] R.N.J. Veldhuis. *Restoration of Lost Samples in Digital Signals.* Prentice Hall, New York, 1990.

[21] R.M. Gray. Vector quantization. *IEEE ASSP Magazine*, 1(2):4–29, 1984.

[22] D. Kleima. *Informatietheorie.* Collegedictaat TU Twente (in Dutch), 1973.

[23] C.E. Shannon. A mathematical theory of communication. *Bell System Tech. J.*, 27:379–423, 623–656, 1948.

[24] D. Huffman. A method for the construction of minimum redundancy codes. *Proceedings of the IRE*, 40(10):1098–1101, September 1952.

[25] J. Ziv and A. Lempel. A universal algorithm for sequential data compression. *IEEE Transactions on Information Theory*, 23(3):337–343, 1977.

[26] J. Ziv and A. Lempel. Compression of individual sequences via variable-rate coding. *IEEE Transactions on Information Theory*, 24(5):530–536, 1978.

[27] T. Bell, I.H. Witten and J.G. Cleary. Modeling for text compression. *ACM Computing Surveys*, 21(4):557–591, December 1989.

[28] J.C. Lawrence. A new universal coding scheme for the binary memoryless source. *IEEE Transactions on Information Theory*, 23(4):466–472, 1977.

[29] J. Rissanen. Universal coding information, prediction and estimation. *IEEE Transactions on Information Theory*, 30(4):629–636, 1984.

[30] N. Farvardin and J.W. Modestino. Optimum quantizer performance for a class of non-gaussian memoryless sources. *IEEE Transactions on Information Theory*, 30(3):485–497, 1984.

[31] S.P. Lipshitz, R.A. Wannamaker and John Vanderkooy. Quantization and dither: A theoretical survey. *Journal of the Audio Engineering Society*, 40(5):355–375, 1992.

[32] J. Max. Quantization for minimum distortion. *IEEE Transactions on Information Theory*, 6:7–12, 1960.

[33] S.P. Lloyd. Least squares quantization in pcm. *IEEE Transactions on Information Theory*, 28:129–137, 1982.

[34] J.C. Kieffer, T.M. Jahns and V.A. Obuljen. New results on optimal entropy-constrained quantization. *IEEE Transactions on Information Theory*, 34(5):1250–1258, 1988.

[35] A.B. Watson. Efficiency of a model human image code. *Journal of the Optical Society of America A*, 4:2401–2417, 1987.

[36] K.W. Cattermole. *Principles of Pulse Code Modulation.* Iliffe Books, London, 1969.

[37] P.F. Panter and W. Dite. Quantization distortion in pulse count modulation with nonuniform spacing of levels. *Proceedings of the IRE*, 39(1):44–48, January 1951.

[38] S.P. Lipshitz, John Vanderkooy and R.A. Wannamaker. Minimally audible noise shaping. *Journal of the Audio Engineering Society*, 39(11):836–852, 1991.

[39] R.A. Wannamaker. Psychoacoustically optimal noise shaping. *Journal of the Audio Engineering Society*, 40(7/8):611–620, 1992.

[40] R.J. van de Plassche and E.C. Dijkmans. A monolithic 16-bit D/A conversion system for digital audio. In B. Blesser, B. Locanthi and T.G. Stockham Jr., editors, *Digital Audio: Collected Papers from the AES premiere conference, Rye, New York, 1982*, pages 54–60. Audio Engineering Society, New York, 1983.

[41] H. Gish and J.N. Pierce. Asymptotically efficient quantizing. *IEEE Transactions on Information Theory*, 14:676–683, 1968.

[42] J.H. Makhoul, S. Roucos and H. Gish. Vector quantization in speech coding. *Proceedings of the IEEE*, 73(11):1551–1588, 1985.

[43] A. Gersho and R.M. Gray. *Vector Quantization and Signal Compression*. Kluwer Academic Publishers, Boston, 1992.

[44] Y. Linde, A. Buzo and R.M. Gray. An algorithm for vector quantizer design. *IEEE Transactions on Communications*, 28(1):84–95, 1980.

[45] B. Juang and A.H. Gray, Jr. Multiple stage vector quantization for speech coding. In *Proceedings ICASSP-82*, pages 597–600, Paris, 1982.

[46] M.J. Sabin and R.M. Gray. Product code vector quantization for speech waveform coding. In *Proceedings IEEE Globecom-82*, pages 1087–1091, Miami, 1982.

[47] A. Buzo, A.H. Gray, Jr. and J.D. Markel. Speech coding based upon vector quantization. *IEEE Transactions on ASSP*, 28(5):562–574, 1980.

[48] IEEE. *Proceedings of the IEEE Workshop on Speech Coding for Telecommunications*, Vancouver, September 1989.

[49] M. Vetterli and D. Le Gall. Perfect reconstruction FIR filter banks: Some properties and factorizations. *IEEE Transactions on ASSP*, 37(7):1057–1071, 1989.

[50] I.I. Hirschmann Jr. Recent developments in the theory of finite Toeplitz operators. In P. Ney and S. Port, editors, *Advances in Probability and Related Topics, Vol. 1*, pages 105–167. Dekker, New York, 1971.

[51] R.N.J. Veldhuis and R.G. van der Waal. Subband coding of stereophonic digital audio signals. In *Proceedings ICASSP-91*, Toronto, 1991.

[52] N. Ahmed and K.R. Rao. *Orthogonal Transforms for Digital Signal Processing*. Springer-Verlag, 1975.

[53] J.D. Johnston. Transform coding of audio signals using perceptual noise criteria. *IEEE Journal on Selected Areas in Communication*, 6(2):314–323, 1988.

[54] R.E. Crochiere and L.R. Rabiner. *Multirate Digital Signal Processing*. Prentice Hall, Englewood Cliffs, New Jersey, 1983.

[55] P.P. Vaidyanathan. Quadrature Mirror Filter Banks, M-Band Extensions and Perfect-Reconstruction Techniques. *IEEE ASSP Magazine*, 4(3):4–20, 1987.

[56] J.W. Woods, editor. *Subband Image Coding*. Kluwer Academic Publishers, Boston, Dordrecht, London, 1991.

[57] J. Makhoul. Linear prediction: A tutorial review. *Proceedings of the IEEE*, 63(4):561–580, 1975.

[58] E. Ordentlich and Y. Shoham. Low-delay code-excited linear-predictive coding of wideband speech at 32 kbps. In *Proceedings ICASSP-91*, pages 9–12, Toronto, 1991.

[59] A. Sugiyama, F. Hazu, M. Iwadare and T Nishitani. Adaptive transform coding with an adaptive block size. In *Proceedings ICASSP-90*, pages 1093–1096, Atlanta, 1990.

[60] R.N.J. Veldhuis, M. Breeuwer and R.G. van der Waal. Subband coding of digital audio signals. *Philips Journal of Research*, 44(2–3):329–343, 1989.

[61] G. Stoll and Y.F. Dehery. High quality audio bit rate reduction system family for different applications. In *Proceedings of Supercomm 90*, Atlanta, 1990.

[62] G.C.P. Lokhoff. DCC - digital compact cassette. *IEEE Transactions on Consumer Electronics*, 37(3):702–706, August 1991.

[63] K. Brandenburg and G. Stoll. The ISO/MPEG–Audio Codec: A generic standard for coding of high-quality audio. *AES Preprint*, (3336), 1992.

[64] E. Zwicker and R. Feldtkeller. *Das Ohr als Nachrichtenempfänger*. S. Hirzel Verlag, Stuttgart, 1967.

[65] E. Zwicker and H. Fastl. *Psychoacoustics*. Springer-Verlag, Berlin, 1990.

[66] L.A. Jeffres. Masking. In J.V. Tobias, editor, *Foundations of Modern Auditory Theory*, pages 87–114. Academic Press, New York, 1970.

[67] M.A. Krasner. The critical band coder. In *Proceedings ICASSP-80*, pages 327–331, Denver, 1980.

[68] G. Theile, M. Link and G. Stoll. Low bit rate coding of high-quality audio signals. *AES Preprint*, (2431), 1987.

[69] J.G. Roederer. *Introduction to the Physics and Psychophysiscs of Music*. Springer-Verlag, New York, 1975.

[70] B. Scharf and S. Buus. Stimulus, physiology, thresholds. In K.R. Boff, L. Kaufman and J.P. Thomas, editors, *Handbook of Perception and Human Performance, Volume 1*, pages 14/1–14/71. John Wiley and Sons, New York, 1986.

[71] R.P. Hellman. Asymmetry of masking between tone and noise. *Perception and Psychophysics*, 11(3):241–246, 1981.

[72] R.N.J. Veldhuis. Bit rates in audio source coding. *IEEE Journal on Selected Areas in Communication*, 10(1):86–96, 1992.

[73] M.J.T. Smith and T.P. Barnwell, III. Exact reconstruction techniques for tree-structured subband coders. *IEEE Transactions on ASSP*, 34(3):434–441, 1986.

[74] E. Zwicker and S. Herla. Über die Addition von Verdeckungseffekten. *Acustica*, 34:89–97, 1975.

[75] G. Lumer. Addition von Mithörschwellen. In *Proceedings DAGA-84*, pages 753–756, 1984.

[76] Y. Shoham and A. Gersho. Efficient bit allocation for an arbitrary set of quantizers. *IEEE Transactions on ASSP*, 36(9):1445–1453, 1988.

[77] MPEG. *MPEG1 Draft Standard. Coding of Moving Pictures and Associated Audio for Digital Storage Media up to about 1.5 Mbit/s*, 1992. Draft International Standard ISO/IEC DIS 11172, ISO/IEC JTC1/SC2/WG11.

[78] J.D. Johnston. Perceptual transform coding of wideband stereo signals. In *Proceedings ICASSP-89*, pages 1993–1996, Glasgow, 1989.

[79] G. Davidson, L. Fielder and M. Antill. High-quality audio transform coding at 128 kbits/s. In *Proceedings ICASSP-90*, pages 1117–1120, Albuquerque, 1990.

[80] M. Idaware, A. Sugiyama, F. Hazu, A. Hirano and T. Nishitani. A 128 kb/s HiFi audio CODEC based on adaptive transform coding with adaptive block size MDCT. *IEEE Journal on Selected Areas in Communication*, 10(1):138–144, 1992.

[81] CCIR (International Radio Consultative Committee). *Encoding Parameters of Digital Television for Studios*, 1982. Recommendation 601-1.

[82] T. Kondo, Y. Shirota, K. Kanota, Y. Fujimori, J. Yonemitsu and M. Uchida. Adaptive dynamic range coding scheme for future consumer digital VTR. In *IERE Proceedings of the 7th International Conference on Video, Audio and Data Recording*, pages 219–226, York, UK, March 1988.

[83] T. Kondo, Y. Fujimori, H. Nakaya, A. Yada, K. Takahashi and M. Uchida. New ADRC for consumer digital VCR. In *IEE Proceedings of the 8th International Conference on Video, Audio and Data Recording*, pages 144–150, Birmingham, UK, April 1990.

[84] A. Habibi. Comparison of nth order DPCM encoder with linear transformation and block coding. *IEEE Transactions on Communications*, 19(6):948–956, 1971.

[85] W. Zschunke. DPCM picture coding with adaptive prediction. *IEEE Transactions on Communications*, 25(11):1295–1302, 1977.

[86] H. Murakami, S. Matsumoto, Y. Hatori and H. Yamamoto. 15/30 Mbit/s universal digital codec using a median predictive coding method. *IEEE Transactions on Communications*, 35(6):637–645, 1987.

[87] J. Salo and Y. Neuvo. A new two-dimensional predictor design for DPCM coding of video signals. In *Proceedings 2nd Int. Workshop on Signal Processing of HDTV*, L'Aquila, Italy, February March 1988.

[88] G. Caronna. Adaptive DPCM with conditional coding. In *Proceedings 3rd Int. Workshop on HDTV*, Torino, Italy, August September 1989.

[89] C.P. Sandbank. *Digital Television*. John Wiley & Sons, Chichester, England, 1990.

[90] P.A. Wintz. Transform picture coding. *Proceedings IEEE*, 60(7):808–819, 1972.

[91] N. Ahmed, T. Natarajan and K.R. Rao. Discrete cosine transform. *IEEE Transactions on Computers*, pages 90–93, 1974.

[92] W.-H. Chen, H. Smith and S. Fralick. A fast computational algorithm for the discrete cosine transform. *IEEE Transactions on Communications*, 25(9):1004–1009, 1977.

[93] M. Breeuwer. Transform coding of images using directionally adaptive vector quantization. In *Proceedings ICASSP-88*, pages 788–791, New York, April 1988.

[94] M. Breeuwer. Full-search versus tree-search vector quantization of discrete cosine transform coefficients. In L. Torres, E. Masgrau and M.A. Lagunas, editors, *Signal Processing 5*, pages 1095–1098, Amsterdam, 1990. Elsevier Science Publishers.

[95] W.-H. Chen and C.H. Smith. Adaptive coding of monochrome and color images. *IEEE Transactions on Communications*, 25(11):1285–1292, 1977.

[96] W.-H. Chen and W.K. Pratt. Scene adaptive coder. *IEEE Transactions on Communications*, 32(3):225–232, 1984.

[97] S. Ericsson. Fixed and adaptive predictors for hybrid predictive/transform coding. *IEEE Transactions on Communications*, 33(12):1291–1302, 1985.

[98] R.C. Reininger and J.D. Gibson. Distribution of two-dimensional DCT coefficients for images. *IEEE Transactions on Communications*, 31(6):835–839, 1983.

[99] G.K. Wallace. The JPEG still picture coding standard. *Communications of the ACM*, 34(4):30–44, 1991.

[100] P.H.N. de With. Motion-adaptive intraframe transform coding of video signals. *Philips Journal of Research*, 44(2-3):345–364, 1989.

[101] P.H.N. de With, M. Breeuwer and P.A.M. van Grinsven. Data compression systems for home-use digital video recording. *IEEE Journal on Selected Areas in Communications*, 10(1):97–121, 1992.

[102] M. Breeuwer. Bit-rate reduction for professional HDTV recording. In *Proceedings of the International Symposium on Fiber Optic Networks and Video Communications*, Berlin, April 1993.

[103] J.W. Woods and S.D. O'Neil. Subband coding of images. *IEEE Transactions on ASSP*, 34(5):1278–1288, 1986.

[104] D. Esteban and C. Galand. Application of quadrature mirror filters to split band voice coding schemes. In *Proceedings ICASSP-77*, pages 191–195, Hartford, 1977.

[105] M. Breeuwer and P.H.N. de With. Source coding of HDTV with compatibility to TV. In M. Kunt, editor, *SPIE Visual Communications and Image Processing '90*, SPIE Processing Series, pages 765–776. SPIE, Bellingham, 1990.

[106] M. Vetterli. Multi-dimensional sub-band coding: Some theory and algorithms. *Signal Processing*, 6:97–112, 1984.

[107] M. Vetterli, J. Kovacevic and D.J. Le Gall. Perfect reconstruction filter banks for HDTV representation and coding. *Image Communication*, 2(3):349–364, 1990.

[108] E. Viscito and J.P. Allebach. The analysis and design of multiresolution FIR perfect reconstruction filter banks for arbitrary lattices. *IEEE Transactions on Circuits and Systems*, 38(1):29–41, 1991.

[109] P.H. Westerink. *Subband Coding of Images*. PhD thesis, Delft University of Technology, October 1989.

[110] P.H. Westerink, J. Biemond and D.E. Boekee. Evaluation of image sub-band coding schemes. In J.L. Lacoume, A. Chehikian, N. Martin and J. Malbos, editors, *Signal Processing 4*, pages 1149–1152, Amsterdam, 1989. Elsevier Science Publishers.

[111] P.H. Westerink, J. Biemond, D.E. Boekee and J.W. Woods. Sub-band coding of images using vector quantization. *IEEE Transactions on Communications*, 36(6):713–719, 1988.

[112] J. Biemond, F. Bosveld and R.L. Lagendijk. Hierarchical subband coding of HDTV in BISDN. In *Proceedings ICASSP-90*, pages 2113–2116, Albuquerque, April 1990.

[113] M. Pecot, P.J. Tortier and Y. Thomas. Compatible coding of television signals, part I: Coding algorithm. *Image Communication*, 2(3):245–258, 1990.

[114] M. Pecot, P.J. Tortier and Y. Thomas. Compatible coding of television signals, part II: Compatible system. *Image Communication*, 2(3):259–268, 1990.

[115] P.J. Burt and E.H. Adelson. The laplacian pyramid as a compact image code. *IEEE Transactions on Communications*, 31(4):532–540, 1983.

[116] F. Kretz and D. Nasse. Digital television: Transmission and coding. *Proceedings IEEE*, 73(4):575–591, 1985.

[117] H.G. Musmann, P. Pirsch and H.-J. Grallert. Advances in picture coding. *Proceedings IEEE*, 73(4):523–548, 1985.

[118] C. Cafforio and F. Rocca. Methods for measuring small displacements of television images. *IEEE Transactions on Information Theory*, 22(5):573–579, 1976.

[119] A.N. Netravali and J.D. Robbins. Motion-compensated television coding: Part i. *Bell System Technical Journal*, 58(3):631–670, 1979.

[120] A.N. Netravali and J.D. Robbins. Motion-compensated television coding: Some new results. *Bell System Technical Journal*, 59(9):1735–1745, 1980.

[121] D.J. LeGall. MPEG: A video compression standard for multimedia applications. *Communications of the ACM*, 34(4):46–58, 1991.

[122] T. Langlais, J.M. Martin and Y. Schifres. TVS 34: Encoding/decoding equipment for 34 Mbit/s TV transmission. *Philips Telecom Review*, 48(3):34–40, 1990.

[123] H.S. Malvar. Lapped transforms for efficient transform/subband coding. *IEEE Transactions on ASSP*, 38(6):969–978, 1990.

[124] O. Rioul and M. Vetterli. Wavelets and signal processing. *IEEE Signal Processing Magazine*, 8(4):14–35, 1991.

[125] M. Breeuwer. Motion-adaptive subband coding of interlaced video. In P. Maragos, editor, *SPIE Visual Communications and Image Processing '92*, SPIE Processing Series, pages 265–275. SPIE, Bellingham, 1992.

[126] W.B. Kleijn. *Analysis-by-Synthesis Speech Coding Based on Relaxed Waveform-Matching Constraints*. PhD thesis, Delft University of Technology, December 1991.

Index

μ-law parameter 65
μ-law quantizer 62, 86

A-law parameter 65
A-law quantizer 62, 86
adaptation 129
adaptation delay 137, 141, 142, 143
adaptation delay of a backward-adaptive source coder 139
adaptation delay of a forward-adaptive source coder 139
adaptation period 132
adaptation rate 129, 132
adaptive bit allocation 130, 144, 148, 157, 165, 222
adaptive dynamic range coding 174
adaptive linear predictive coder 130
adaptive linear predictive coding 143
adaptive pulse-code modulation 19, 174, 175
adaptive quantizer 25
adaptive source coding 129
adaptive source-coding system 16, 19
adaptive subband coding 141
adaptive transform coding 140
adaptive zonal coding 202
adaptivity 129
additive error model 52
additive masking 163
additive source-coding error 11
ADPCM 125, 126
ADRC 174
algorithmic delay 132
alias component 214
aliasing 225
aliasing in subband coding 214
allocation window 166
alphabet 34
analog-to-digital converter 7
analysis-by-synthesis method 126
APCM 19, 22, 24, 30, 83, 97, 130, 132, 134, 135, 136, 139, 141, 143, 149, 157, 161, 174, 175
arithmetic coding 44
autocorrelation function 90, 105
autocorrelation matrix 90, 107
autoregressive process 94
average codeword length 33, 36, 37
average entropy per quantized subband sample 115
average entropy per sample 105
average entropy per transform coefficient 105, 109
average number of bits per symbol 38, 47

backward adaptation 143
backward-adaptive coder 138
backward adaptivity 129, 130
bark 159
bark rate 163
bark scale 163
binary code 34
binary codeword 21, 25, 32
binary prefix-condition code 35, 37
binary string 33
binary tree 35
binary tree-search VQ 83
bit allocation 177, 201, 202, 221
bit allocation table 203
bit rate 8, 51, 70, 85, 88, 99, 112, 115, 116, 118, 119
bit-rate control 19, 130, 144, 192
bit-rate control block 15, 16
bit rate of digital pictures 172
bit rate of high-definition TV 172
bit rate of standard-definition TV 172
block-companded coding 19
block-companded quantizer 86
block companding 19
blocking artefacts 223
blocks of a source-coding system 15, 19
buffer 144
buffer control 130, 144, 192

buffer overflow 145
buffer regulation 192, 204
buffer status 145
buffer underflow 145

CCIR 601 TV signal 172, 174, 178, 188, 222, 232, 239
CELP 126
central limit theorem 121
channel 3
channel error 4
chrominance 170
closed-loop prediction 181
codebook 72, 126
codebook design 79
codebook-excited linear predictive coders 126
codebook vector 73
codec 5
coder 51, 87
coder block 24, 25
coder complexity 89, 109, 117
codes 34
codeword length 34
coding 173
coding block 15, 16, 19, 32
coding delay 129, 137, 141, 142, 143
coding error 50, 156
coding-error correlation matrix 98, 111
coding-error variance 99, 119
coding of discrete sources 32
colour picture 170
combination of a quantizer and a coder 51, 68
compact disc signal 156
companding quantizer 65
compatible coding of HDTV 223, 231
complexity 14
compression function 65
conditional entropy 46
conditional probability 45
conjugate quadrature-mirror filter bank 160
conjugate quadrature mirror filters 215
correlation matrix 100, 104
CQF 215
CQF filter banks 113, 160
critical band 159
critical-band noise target 159
critical-band rate 159
critical bandwidth 159

DCC 155, 156
DCT 87, 99, 108, 192, 194, 195, 237
DCT basis matrices 195

dead-band uniform quantization 204
dead-zone uniform quantization 204
decimating filter bank 113, 160
decimation 113
decision level 28, 51, 56, 59, 61
decision threshold 28, 51
decoder 87
decoder block 24, 25
decoding block 19
decoding delay 142
decomposition into basis vectors 110
delay 11, 14, 129, 130, 132, 137
demodulator 3, 4
destination 3, 5
DFD 236
DFT 99, 195
differential entropy 69, 70
differential entropy of a vector source 90
differential pulse-code modulation 9, 119, 173, 181, 233
digital-to-analog converter 7
digital broadcasting 156
Digital Compact Cassette 156
digital picture 51, 169
discrete-value continuous-time signal 7
discrete-value discrete-time signal 7
discrete cosine transform 87, 99, 108, 111, 168, 192, 194, 195, 237
discrete Fourier transform 99, 111, 168, 195
discrete source 32, 51
displaced-frame difference 236
distortion 4, 8, 9, 10, 19, 28, 50, 68, 70, 87, 120, 150
distortion measure 10, 28, 68, 70, 89, 99, 119
dither 60, 65
DPCM 9, 24, 119, 123, 173, 181, 233
dynamic range 175

effective parameter-estimation memory size 137
eigenvalue 91, 105
eigenvalues of the autocorrelation matrix 91
eigenvector 91, 105
electronic still picture camera 185
emergency measure 145
encoding delay 141, 142
end-of-block 206
entropy 33, 36, 37, 38, 45, 46, 47, 68, 88, 103, 112, 115, 118, 119, 120, 187
entropy-optimized quantizer 62, 85
entropy of a quantized signal 88
entropy of quantized transform coefficients 200

Index   261

entropy of the quantizer-output symbols 52, 56, 68
entropy of a sequence 45
entropy per symbol 45, 91
EOB 206
error correction 3, 4
errorless source coding 8
error protection 3, 4
ESP 185
excitation sequence 126
excitation vector 126
expansion function 65

field 207
filter bank 211, 215
fixed bit rate 8, 19, 144, 157
fixed-length coding 25, 43, 73, 85, 176
fixed-to-fixed length coding 43
fixed-to-variable length coding 44
FLC 43, 176
forward-adaptive coder 138
forward-adaptive prediction 183
forward adaptivity 129, 130
Fourier transform 170, 224
Fourier transform of subband signals 212
frame 170, 207
frame rate 170
frequency bands 211
frequency-dependent quantization 196, 198, 202
full-search block matching 236
full-search vector quantizer 74

G.722 speech coder 142
gain-shape VQ 81, 83
Gaussian joint probability density function 90, 100
Gaussian probability density function 53, 54, 62, 70
Gaussian signal 90
Gish-Pierce asymptote 71, 72, 85, 94
granular noise 53, 54

HDTV 172, 231
high-definition television 172, 231
high-rate quantization 68
histogram 187, 232
Huffman coding 26, 39, 44, 48, 70, 72, 88, 89, 103, 106, 117, 168, 187
human observer 5, 111
human visual system 174, 175, 186, 192, 194, 198, 211, 220, 222, 229

implementation delay 132

impulse response 212
in-band masking 157, 163
index 51, 73
integer-value discrete-time signal 6
interframe coding 232
interlaced TV signals 207
interpolating filter bank 113, 160
interpolation 113
intraframe coding 232
inverse-transformation block 19, 24, 28
irrelevancy 9, 19, 27, 28, 173
ISO 168

Joint Picture Experts Group 206
JPEG 206

Karhunen-Loève transform 107, 108, 109
KLT 107
Kraft 35
Kraft's inequality 36

Laplacian probability density function 57, 61
lattice filter 123, 143
LBG algorithm 80, 81
Linde-Buzo-Gray algorithm 80
linear prediction 122, 126
linear predictive coder 130
linear predictive coding 119
logarithmic noise-to-mask ratio 167
logarithmic quantizer 62, 166
lossless data compression 44
lossless source coding 8
LPC 119
luminance 9, 169

Macmillan 36
magnetic video recording 178, 207
Markov process 48
masked power 157, 161, 163, 166
masker 156
masking 156, 157
masking model for subband coding 162
masking threshold 156, 158
Max-Lloyd quantization of subband signals 221
Max-Lloyd quantization of transform coefficients 201
Max-Lloyd quantizer 61, 72, 75
mean-square error 11, 28, 53, 68, 89, 98
measure of information 39
median prediction 183
medical picture coding 231
memoryless quantizer 50

message 34
midrise quantizer 56, 85
midtread quantizer 56
midtread uniform quantizer 66
modulator 3
motion-compensated prediction 235
motion estimation 236
Motion Picture Experts Group 239
motion vector 236
MPEG 239
MSE 11
multi-dimensional signal 7
multiple noise target 162
music coding 142

noise-shaping 28, 53, 60, 111, 112
noise-shaping filter 65, 121, 125
noise-shaping quantizer 51, 65, 119
noise target 158, 162
noise-to-mask ratio 167
noise-weighting filter 126
non-recursive parameter estimation 134
non-stationary sequences 46
non-subtractive dither 60
non-uniform quantizer 61, 69, 85

objective quality 5
one-dimensional signal 7
optimal prefix-condition code 39
orthogonal transform 29, 87, 96
out-of-band masking 157, 163
overlapping transforms 168
overload distortion 53, 70, 103, 135
overload error 53, 54, 66

parameter coding 11
parameter estimation 129, 130, 133
parameter-estimation delay 132
parameter-estimation memory 136
PASC 156, 161, 165
passband 212
PC 173, 224, 227, 230
PCM 43, 173, 174, 181
pdf-optimized quantizer 54, 61, 62, 85
pdf-optimized uniform quantizer 62
peak value 20, 28, 130, 161
pel 170
perception 110, 118, 125
perceptual coding 112
perceptual distortion 111
perceptually optimized quantizer 62
perceptual spectral weighting 119
perceptual weighting of the coding error 126

picture 169
picture coding 99, 108, 146, 169
picture element 169
picture rate 170
picture sequence 170
picture source coding 169
pixel 9, 170
polyphase filter banks 113
positive definite matrix 91
power concentration 108, 109, 113, 116
power spectral density 10, 90, 92, 143, 158, 196, 229
power spectral density of the quantization error 66
Precision Adaptive Subband Coding 156
prediction 181, 183, 233, 235
prediction coefficients 130, 183
prediction error 119, 181, 187, 235, 236
predictive coding 88, 141, 178, 181
predictive filtering 29, 87, 119
prefix 187, 206
prefix condition 35
prefix-condition code 38, 39, 47
probability density function 10, 28, 33
probability of occurrence 32
pulse-code modulation 43, 173, 174, 181
pyramidal coding 173, 224, 227, 230

QMF 214
QMF filter banks 113, 160
quadrature-mirror filter 214
quadrature-mirror filter bank 160
quantization 32, 50, 173, 237
quantization and coding 68
quantization block 15, 16, 19, 20, 24, 27, 145
quantization characteristic 21, 28, 52, 60, 64, 68, 185, 197, 201
quantization error 28, 53, 60, 66, 97, 177, 185, 197, 218, 227
quantization-error variance 53, 54, 60, 61
quantization in a predictive coder 119
quantization in pyramidal coding 227
quantization level 20, 27, 56, 59, 61, 177
quantization noise 83, 196
quantization step size 177
quantization of subband signals 217
quantization of transform coefficients 196, 197
quantizer 10, 20, 27, 50, 87
quantizer-coder combination 51
quantizer-input value 20, 27, 51
quantizer-output symbol 20, 21, 25, 27, 32, 50, 51, 87

Index

quantizer-reconstructor 52
quantizer as a discrete source 50, 51
quantizer as an error source 50, 52
quantizer with memory 50, 65
quantizing and coding of dependent sources 88

random process 32
rate-distortion bound 12, 71, 72, 86, 93
rate-distortion function 12
rate-distortion theory 12, 70, 93, 110
rate-distortion theory for Gaussian signals 51
rational power spectral density 92
real-value continuous-time signal 5
real-value discrete-time signal 6
reconstruction block 19, 27
reconstruction error 197
reconstructor 20, 50, 87
reconstructor block 24
reconstructor-output value 50
recursive parameter estimation 134
reduction factor 178, 222, 237
redundancy 8, 19, 27, 28, 39, 48, 50, 173, 192, 232
ringing 221
ringing artefacts 223
runlength coding 42, 44, 48
runlength coding of transform coefficients 205

sampling raster 169
SBC 173, 211, 229, 230
scalar quantizer 27, 51
scanning of transform coefficients 205
scanning pattern 205
SDTV 172, 231
second-order stationary signal 129
sensitivity to channel errors 15
sensitivity to model deviations 15
Shannon 34
Shannon's source-coding theorem 36, 37, 38, 47
short-time music power spectral density 162
side information 175
signal model 12
signal parameters 130
signal-to-noise ratio 11, 29, 54, 62, 83, 161
signal-to-quantization-noise ratio 220
simultaneous masking 156
SNR 11, 220
SNR-optimized quantizer 62, 85
sound-pressure level 157, 163
source 3, 5

source coding 3
source-coding error 29
source-coding system 32, 87
source decoder 3, 4, 19, 24
source-decoder delay 132
source encoder 3, 4, 19, 24
source-encoder delay 132
source signal 5
spatial frequency 171, 192, 216
speech coding 119, 142, 143
SPL 157
standard-definition television 172, 231
stationary process 33
stationary signal 11, 16, 129
statistical distortion measure 11
statistically dependent symbols 33, 45
statistically independent symbols 33
statistical signal model 12
statistics of transform coefficients 205
step size 57, 59, 88
stochastic process 32
stopband 212
subband coding 88, 115, 141, 157, 160, 173, 211, 229, 230
subband coding of audio signals 156
subband filtering 29, 87, 113
subband filtering of pictures 216
subjective quality 5
suboptimal transforms 108
subtractive dither 60
symmetrical quantizer 56
Szegö limit theorem 92

target 156, 158
TC 173, 192, 229
television 172
temporal frequency 171
threshold coding 203, 205
threshold of hearing 157
threshold sampling 203
Toeplitz matrix 90
tonal masker 163
training procedure 79
transfer function of subband coding 214
transformation 28, 87, 173
transformation block 15, 16, 19, 24, 28, 52
transform basis matrices 194
transform coding 88, 110, 140, 168, 173, 192, 229, 237
transform coefficients 192, 194
transmission rate 8
transmission system 3
tree-searched VQ 81

TV 172

uniform midtread quantizer 100
uniform probability density function 53
uniform quantization 229
uniform-quantization noise 60
uniform quantization of subband signals 220
uniform quantization of transform coefficients 201
uniform quantizer 20, 59, 85, 88, 161
uniquely decodable binary code 38, 47
uniquely decodable code 34
uniquely decodable codewords 33
uniquely decodable message 34
universal coding 44

VAPCM 83
variable-length coding 26, 44, 72, 85, 168, 183, 192, 235
variable-length coding of transform coefficients 205
variable-to-fixed length coding 44
variable-to-variable length coding 44
variable bit rate 8, 144
vector APCM 83
vector quantization 72, 126
vector quantization of transform coefficients 201
vector quantizer 27, 51, 72, 86, 143
videoconferencing 238
videophone 172, 238
VLC 44, 183, 192, 235
VQ 74, 81

Walsh-Hadamard transform 99
waveform coding 11
weighted mean-square error 126
weighting factor 198, 202, 222
weighting function 200
weighting of transform coefficients 196
WHT 99
wide-sense stationary signal 129

zero-mean additive white noise 60
zig-zag scanning pattern 205
Ziv-Lempel coding 43, 44, 129
Ziv-Lempel-Welch procedure 43
zonal coding 202
zonal sampling 202